COMPARATIVE ENVIRONMENTAL POLITICS

ADVANCES IN GLOBAL CHANGE RESEARCH

VOLUME 25

The titles published in this series are listed at the end of this volume.

COMPARATIVE ENVIRONMENTAL POLITICS

By

Jerry McBeath

University of Alaska Fairbanks,
Department of Political Science, U.S.A.

and

Jonathan Rosenberg

University of Alaska Fairbanks,
Department of Political Science, U.S.A.

A C.I.P. Catalogue record for this book is available from the Library of Congress.

ISBN 13 978-90-481-7187-3
ISBN 10 1-4020-4763-0 (e-book)
ISBN 13 978-1-4020-4763-3 (e-book)

Published by Springer,
P.O. Box 17, 3300 AA Dordrecht, The Netherlands.

www.springer.com

Printed on acid-free paper

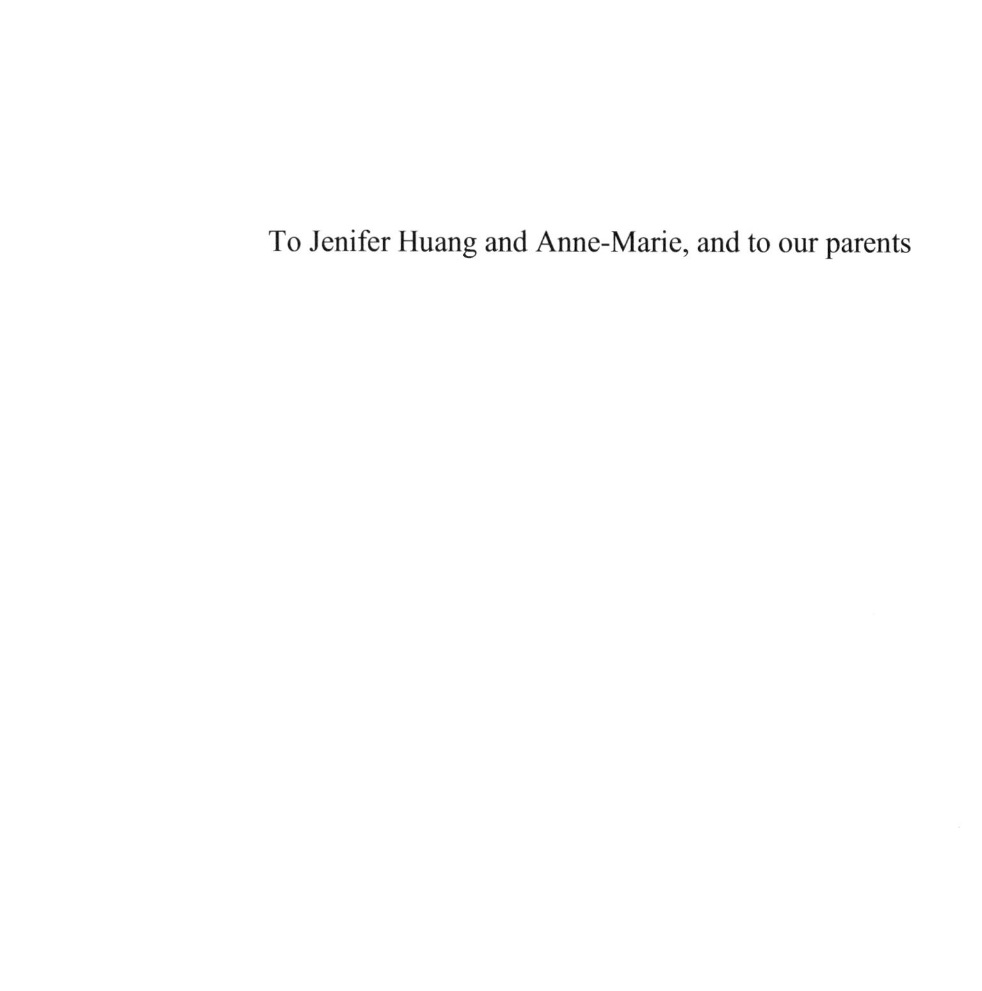

To Jenifer Huang and Anne-Marie, and to our parents

CONTENTS

ACKNOWLEDGMENTS

We began work on this volume in 2001, by preparing documents for a new course in the curriculum on Global Environmental Policy at the University of Alaska. We thank Joe Kan, then dean of UAF's Graduate School, for funding course preparation costs. Since then, three cohorts of students have taken Comparative Environmental Politics, and we appreciate their constructive criticism of the text. Graduate assistants Francine Jenness, Rockford Weber, and Andrew Quainton helped chase down voluminous materials, for which we are grateful.

Several agencies and academic institutions supported our research on environmental politics during this period. NSF/CIFAR supported McBeath's studies of human responses to climate warming; NOAA/CIFAR funded his research on scientific uncertainty concerning fisheries and endangered marine species; and EPA's global change program aided his study of traditional ecological knowledge of toxic contamination and trans-boundary pollution. A UAF sabbatical and Fulbright grant enabled him to conduct research on environmental politics in China, and a Chiang Ching-kuo award extended study to Taiwan. Both the China Foreign Affairs University in Beijing and National Chengchi University in Taipei were gracious hosts. NSF supported Rosenberg's research on sustainable development in the Caribbean, and a sabbatical facilitated broader studies of Caribbean and Latin American environmental topics. Dr. Harvey Feigenbaum of the Elliott School of International Affairs at George Washington University, Dr. Calum Macpherson of the Windward Islands Research and Education Foundation of St. George's University in Grenada, and Dr. Fernando Lopez-Alves and the staff of the Universidad Abierta Interamericana in Buenos Aires provided generous in-kind support for Rosenberg's sabbatical research. We are deeply thankful for the confidence funding agencies and our colleagues at various institutions reposed in us, and hold them harmless from errors and omissions that might remain in the text.

We thank our colleagues and particularly Jim Gladden, Amy Lovecraft, and Karen Erickson, for their congeniality, collegiality, and intellectual stimulation. They were ready sounding boards for our ideas and willingly shared insights, helping us think more reflectively about the sub-disciplines of political science and their interactions with environmental politics. Also, we thank the administrative assistants and student workers in political science, women's studies, history, and northern studies, and especially Courtney Pagh, Julia Parzick, Jaime Kearney and Alicea McCoy for assistance in ways too numerous to mention.

Scholarly undertakings intrude on family time, and we thank our loving kin—Jenifer Huang, Bowen, and Rowena McBeath; Anne-Marie Poole, Abraham, Estelle and Ruth Ann Rosenberg—for tolerating frequent absences and missed family occasions. They encouraged our research and study, raised our spirits when they flagged, and also cut us some slack. We are in their debt.

Finally, we thank Margaret Deignan of Springer for assistance throughout the project, and especially for her understanding and patience.

ABOUT THE AUTHORS

Jerry McBeath is professor of political science at the University of Alaska Fairbanks, where he has taught since 1976. He was educated at the University of Chicago (BA, 1963; MA in international relations, 1964) and at the University of California, Berkeley (Ph.D., political science, 1970). Although his primary research specialization is comparative politics (with a focus on East Asia and emphasis areas in studies of China and Taiwan), his publications also include books and articles on Alaska state and local government, the Alaska Constitution, Alaska Native politics, rural Alaska education, education reform in the American states, government and politics of circumpolar northern nations, and American and European environmental politics.

Jonathan Rosenberg is associate professor of political science at the University of Alaska Fairbanks where he joined the faculty in 1993. He holds a BA in political science from the Pennsylvania State University (1980) and a Ph.D. in political science from the University of California, Los Angeles (1992). His research interests span comparative politics and international political economy with a concentration in Latin America and the Caribbean. His publications include articles and chapters on the role of development assistance in participatory resource management, stakeholder participation and sustainable development in the Eastern Caribbean, and Cuban and Mexican political economy.

CHAPTER 1. INTRODUCTION

1. THE NATURE OF THE SUBJECT

1.1 Crises and Hope

On the night of December 2, 1984 a chemical plant owned by the US-based multinational corporation Union Carbide spewed approximately 27 tons of deadly methyl isocynate gas into the air around Bhopal, India. An estimated 500,000 people were exposed; officially 15,000 deaths have been attributed directly to exposure, with thousands of others suffering long-term damage to their health.[1] Local flora and fauna were seriously damaged.

On March 23, 1989 the Exxon Valdez received its cargo at the terminus of the 800-mile Trans-Alaska Pipeline in the port of Valdez, on Alaska's Prince William Sound. A little past midnight on March 24, the tanker ran aground on Bligh Reef, releasing 10.8 million gallons of Alaskan crude oil into the icy waters and rocky shoreline of the sound.[2] Fragile marine habitat was polluted and untold numbers of marine mammals, fish and sea birds were killed.

On October 16, 1989 municipal authorities and the federal Environment Minister declared an air pollution emergency for Mexico City. Ozone levels had reached three times the maximum deemed acceptable by the World Health Organization. One third of the city's motor vehicles were ordered off the streets; factories were mandated to cut production to 30 percent of normal; schoolchildren were forbidden to play outdoors; the elderly and people with respiratory ailments were advised to stay indoors and avoid physical activity.[3]

These three events, part of a seemingly relentless litany of bad environmental news accumulating over the last three or more decades, can easily lead to despair and disillusionment. But what do they actually mean for the health of the earth and its inhabitants? When the political implications and reactions to incidents like these are examined the picture that emerges is more complex, and possibly more hopeful, than a mere linear dissent into impending doom.

Most environmental disasters, despite their localized effects, result from complex mixtures of global, national and local factors, and therefore can be affected by a wide range of governmental and non-governmental actors at several levels of governance simultaneously. Incidents like the ones described above elicit broad responses, including new government policies at the local and national levels, intensified activities by local and international non-governmental organizations, and new corporate policies and strategies. But the types and effectiveness of responses seem to vary considerably from incident to incident and country to country.

In Bhopal, a new agency was established by the state of Madhya Pradesh for the exclusive purpose of dealing with the aftermath of the tragedy, administering new state and federal regulations, and coordinating the relief efforts of state, federal and non-governmental agencies. Litigation led to multi-million dollar settlements. Union Carbide paid nearly US$500 million to the Government of India. But the actual indemnification of victims remains inadequate and incomplete, and victims continue their often frustrating efforts to find remedy through the U.S. and Indian legal systems.[4]

In Alaska, Exxon faced criminal charges and civil suits, eventually paying US$25 million in indemnification to the State of Alaska, and providing US$900 million for habitat restoration. The US Congress passed the 1990 Oil Pollution Act applying new regulations to the design and operation of oil tankers, the state passed new oil spill legislation, and state and federal governments established the Exxon Valdez Oil Spill Trusteeship Council to administer the restoration funds.[5]

In Mexico City some progress had been made toward mitigating the polluting effects of choking automobile traffic and industrial activity in the valley. The federal government pledged US$13.3 million for pollution reduction programs between 1996 and 2000; new federal regulations, tax incentives and subsidies promoted cleaner, more fuel efficient vehicles;[6] and the mayor of Mexico City proposed changes to the city's mass transit and road systems aimed at improving traffic flows and reducing the use of private passenger cars and small buses. In 2002, the Global Environment Facility (through the World Bank) granted US$6.125 million to help improve Mexico City's air quality.[7] Nevertheless, Mexico City remains one of the most polluted urban environments in the world.

These three incidents have their unique elements, but they are also comparable. Each one has generated intensive negative publicity, has raised awareness of the pervasiveness of environmental problems, and has led to outpourings of aid, sympathy, indignation and defensiveness. Each disaster led to litigation, the promulgation of new policies, and the creation of new institutions. But the effectiveness of the responses has varied greatly. So how do we understand the similarities in the types and quality of the responses? Is it the nature of the environmental problems—air-borne vs. marine pollution? Or is it a function of the socio-political differences of the three countries—levels of development, institutional structures and capacity—which best explain the variability?

While we cannot hope to provide definitive answers to these questions, as political scientists we proceed from the assumption that much can be learned through the application of the existing tools of comparative political analysis. Two of the incidents occurred in developing countries (one a parliamentary democracy, the other a democratizing presidential system), and one occurred in a highly developed presidential democracy. All three countries are federal systems, with multiple layers of environmental regulation, although sub-national government seems to be better institutionalized and more effective in the United States. Only one of the incidents occurred within the sovereign territory of a country with an effective federal environmental protection agency and highly transparent policy and judicial processes. None of the three nation-states involved has a strong "green" political party. And the affected communities in all three cases remain dissatisfied with the outcomes to varying degrees. What can we make of these observations? Is there a systematic relationship between them (or any other political, social, cultural or economic factors) and the responses to environmental crises by nation-states? To answer these questions a more systematic analysis of the attendant political processes is needed. Only then can we suggest whether hope or despair is a more appropriate reaction.

1.2 Global Environmental Issues

Since the 1960s, environmental issues have entered the agendas of most nation-states. Pollution of land, air, and water have endangered ecosystems and public health, and called for a governmental response. Problems of water scarcity and depletion of other critical natural resources, such as forest and agricultural land, elevated the salience of environmental issues, as did the incessant accumulation of garbage. These issues introduced new sets of problems to the political arena and brought new sources of demands to bear on governments. In the 1970s and 1980s governments, especially those of the economically developed countries (EDCs), created new institutions, such as environmental protection bureaus, to resolve environmental problems. By the late 1980s, such institutions were nearly universal. Around the same time environmental movements were becoming more active and articulate in their pursuit of policy goals, and political parties devoted to environmental issues and environmentalist perspectives were forming and beginning to contest elections.

Although the global nature of environmental problems was acknowledged at the 1972 United Nations Conference on the Human Environment, until the 1980s, environmental problems were thought to be susceptible to national solution, because they appeared to occur primarily within the territorial confines of states and could be addressed with existing forms of administration. To a certain extent the reliance on national solutions was a matter of making a virtue out of necessity as environmental problems were not seen as urgent enough to warrant nation-states relinquishing their sovereignty in search of solutions. Where cooperation among states did exist it was carried out within the context of the cold war. For the great powers, resource and development issues were largely connected to national security considerations, and the environment took a back seat. Since that time, however, a series of environmental issues with global implications has drawn the attention of scientists, policy-makers, the media, and the mass public:

- Climate warming, caused by natural factors as well as the dramatic increase in greenhouse gas emissions since the industrial revolution;
- Biological diversity loss, caused by pressures of logging, agriculture, housing, and commercial development on the critical habitat of endangered and threatened species;
- Deforestation, particularly of tropical forests, as a result of increased and unsustainable logging (both legal and illegal);
- Desertification, due to natural erosion and drought as well as deforestation and agricultural/commercial development in areas with marginal soils;
- Trans-boundary air pollution, including acid rain and persistent organic pollutants (POPs), caused largely by emissions from industrial sites; and
- Pollution of the world's oceans, and depletion of ocean resources, especially fisheries.

The development of global environmental issues and prognoses of impending environmental disasters have had two effects. First, they have energized attempts to create mechanisms of environmental governance at the international level, and the past two decades have witnessed a virtual explosion in the number of international environmental conferences and conventions to mitigate environmental problems. Second, the fact that each of these global environmental issues has domestic as well as foreign origins, and that nation-states claim sovereignty and control over domestic issues, has directed the search for solutions to individual countries.

1.3 Comparative Environmental Politics

Comparative politics is a sub-field of political science, which examines primarily the national (and sub-national) structures of countries, their political processes and values. Scholars in this sub-field may compare one country to a model or pattern; they may compare a small number of countries, either with different or similar attributes[8]; or they may compare a large number of nations, perhaps all, which implies the use of quantitative methods of analysis.[9] Whatever set of countries is examined, the objective of comparative politics is to understand and explain the outputs and outcomes of state behavior (for example, degree of civil liberties protection or amount of schooling delivered). The comparison process tells us whether similar policy outcomes are the product of similar or different structural and behavioral arrangements within nation-states, and whether the same kinds of power arrangements produce similar or different results.

Comparative *environmental* politics focuses on national and sub-national differences and similarities in environmental policy and environmental outcomes, and attempts to explain their origin. It is thus a relatively specialized subset of national policies (and influencing variables) concerning the totality of the physical conditions in which the nation-state and its people live. Unlike most other policy fields, comparative environmental politics is particularly reliant upon knowledge produced in the biological and other natural sciences concerning ecosystem, plant, and animal changes. But similar to other policy areas (for example, educational, economic, and health policy) environmental policy debates tend to politicize scientific data, research and disagreements. In other words, comparative environmental politics concerns not only the interplay of competing interests in articulating and solving environmental problems, but the production and use of scientific knowledge to be applied to policy decisions and institutional design.

2. THE RELEVANCE OF COMPARATIVE ENVIRONMENTAL POLITICS

As has been pointed out frequently, environmental problems tend to be global and transboundary by nature. As such, they challenge the capacity of nation-states to make and implement effective policy. Environmental issues and environmental problem solving have figured prominently in discussions of globalization, where they are frequently cited as factors in the declining relevancy of nation-states, assaults on sovereignty, and the rising importance of transnational governmental, non-governmental and commercial actors. For political scientists, debates among globalization scholars about the relevance of the nation-state in international politics are also debates about the relevance of comparative political analysis for understanding the world.

Globalization clearly challenges old assumptions about an international system of self-interested, self-motivated, and largely self-contained states. International relations scholars identify several important changes under the rubric of globalization: increasing international trade and investment; declining numbers of wars between states (and increasing incidents of state-less terrorism, intra-state and inter-communal violence); technologically driven explosions in transportation and communication; growing international political networks; standardization of beliefs about political and economic systems (i.e., a global preference for democracy and free markets); increased importance for international organizations such as the United Nations and World Bank, and treaty organization such as the North American Free Trade Agreement and the World Trade Organization; regional integration, especially the European Union; the awesome power of private transnational actors (including multi-national corporations and non-governmental organizations); and the homogenization of popular culture in what might be called the jeans, tee-shirt, running shoes, English-language, and hip-hop phenomenon.[10]

Debates about the effects of globalization on states are taking place on two levels. First, there is the debate about sovereignty. International relations scholars disagree about the present and future of the sovereign nation-state. James N. Rosenau argues:

> The very notion of 'international relations' seems obsolete in the face of an apparent trend in which more and more of the interactions that sustain world politics unfold without the direct involvement of nations or states.[11]

Steven D. Krasner, on the other hand, declares that "the most important impact of economic globalization and transnational norms will be to alter the scope of state authority rather than to generate some fundamentally new way to organize political life."[12]

Neither side is completely convincing. As Krasner also points out, sovereignty has never been as powerful or absolute a factor in international relations as its proponents would like or its critics fear. And a state's sovereignty tends to vary positively with its level of development. Powerful, wealthy states with capable institutions are more sovereign than weak, poor and politically unstable states. This observation in itself points out the value of broad, comparative political analysis for understanding the effects of globalization.

Second, there is the debate about domestic and international influences on policy making. Globalization has taken the old "levels of analysis" controversy in international relations theory and stood it on its head.[13] Scholars now argue, not only over the relative importance of international and domestic determinants of foreign policy but of domestic policy as well. For example, are changes in the enforcement practices of the U.S. Environmental Protection Agency under the George W. Bush administration a response to competitive pressures from a globalizing economy or political pressures from domestic interest groups?[14]

Given the preoccupation of scholars, policy-makers and activists with globalization, it is not surprising that the bulk of the political science literature on world environmental politics comes from the sub-fields of international relations and international political economy. But what becomes clear in perusing this literature is the need for a better understanding of the roles of domestic social forces and the political structures of nation-states. In their recent edited volume, *the Global Environment*, Axelrod, Downie, and Vig devote the first two sections to international institutions and global policy questions, but reserve a third and final section for analyzing global policies on sustainable development at the national and EU-levels because,

> the concept of sustainable development is quite broad and has quite different meanings when translated into different cultures and languages. . . Some nations such as New Zealand and the Netherlands have adopted far-reaching sustainable development plans and programs, whereas others have dealt with sustainability issues in a piecemeal and ad hoc fashion, if at all.[15]

To know why, we have to know more about these states, their societies, histories and cultures. Therefore, in this book we take the position that nation-states and their governments still matter for three reasons. First, they are the locus of decision-making for a wide range of economic, social, cultural and resource management policies that affect the global environment, National governments, then, are the prime targets of local, national and transnational environmental activism. Second, only national governments can decide whether to join or not join, cooperate or not cooperate with international environmental agreements, treaties and protocols. And finally, many of the differences we find among the environmental policies and situations of nation-states depend on domestic political variables, including ideology, regime type, political culture, state-society relations, and scientific and institutional capacity.

A comparative approach to policy illuminates well the different stances adopted by nation-states regarding global environmental problems such as climate change. Take, for example, two of the leading industrial powers of the world—the United States and Germany. Scientists in both nations have confirmed the dramatic increase in greenhouse gas emissions through the twentieth and into the twenty-first centuries. With the exception of a minority of "greenhouse skeptics," climate scientists link the rise of temperatures, particularly evident in polar regions, to the increase of greenhouse gas emissions. They do not attribute it primarily to natural climate cycles or other factors.

One might expect the two nations to have developed similar policies toward the mitigation of climate warming, given their comparable degree of modernization and level of economic development. Indeed, both joined the Framework Convention on Climate Change and participated actively in negotiations leading to the 1997 Kyoto Protocol. However, Germany ratified the protocol and quickly proceeded to implement its provisions, while

President Clinton, whose administration negotiated the treaty, did not submit it to the U.S. Senate for ratification. Then, within one year of his inauguration to the presidency, George W. Bush removed the United States from the influence of the protocol.

Notwithstanding similar economic and social systems, it was differences in political institutions and processes that explained the divergent environmental outcomes. Germany has one of the world's largest and most active Green parties, and it became an attractive coalition target for the Social Democrats when they attained the largest number of seats in the Bundestag elections of 1998. The price of Green support for a Social Democratic government was two ministries—Environment and Foreign Affairs—as well as policy stances in accord with several important planks of the Green platform (reduction of greenhouse gases and even eco-taxes).

Differences in electoral institutions and degree of concentration in decision-making authority created less auspicious conditions for change in the United States. The American single-member district, plurality election system discourages formation and electoral success of third parties such as the Greens. At its contemporary high point in the 2000 presidential election, Green Party candidate Ralph Nader won no state's electoral votes and only 2.7 percent of the popular vote. Although in the estimation of Democratic partisans, this was sufficient to spoil the chances of Democratic presidential candidate Al Gore, it was far from enough support to influence national policy. Moreover, the U.S. separation-of-powers system that divides national power among the presidency, Congress, and the courts would have made it improbable for Al Gore, had he won the Electoral College vote, to have won a two-thirds majority vote for the Kyoto Protocol in the U.S. Senate, which was evenly divided after the 2000 election.

Differences in political organizations and state-society relationships also help explain the differences between climate change positions of Germany and the United States. Germany produces little of its own energy needs, and the oil and gas industry plays a relatively weak role in national politics. In the German corporatist system economic interests may more easily influence state policy than in the United States (if the executive and legislative branches are unified). Yet interests of single sectors, such as the oil and gas industry, are considerably weaker than in pluralist systems like the United States. There, well-financed political action committees (PACs) of major corporations may influence elections of members of Congress and presidents through campaign contributions and lobbying activities, particularly if they form a broad coalition such as the Climate Change Coalition, which was opposed to U.S. ratification of the Kyoto Protocol. This influence extended to media reporting on the climate change debate, which largely echoed the coalition's skeptical position on the linkage between carbon dioxide emissions and climate warming.

Thus, the analysis possible through comparative politics explains the variation we see in national positions on global climate change. It enlarges our ability to understand complex events and somewhat confusing processes.

3. UNIT OF ANALYSIS: THE NATION-STATE

Although we will discuss differences among and between countries because of their territorial distribution of power (for instance whether they are unitary or federal, and what this implies for the behavior of sub-national governments), our primary focus of interest is the nation-state. Nearly 360 years after the Treaty of Westphalia acknowledged state sovereignty and equality as the key principles of international relations, nation-states remain the primary actors in world politics. In 2006, there are 192 nation-states, and they express great variation in size of territory and population, military power, culture, society, and wealth—as well as in political system characteristics.

A small number of states are territorially vast and cover many climate zones and ecosystem types, such as Russia, Canada, China, and the United States. At the opposite pole are the world's micro-states such as Vanuatu, Palau, Monaco, which have less land than the average European city or Indian village. A few states have huge populations; China, with 1.3 billion and India, with 1 billion people, together comprise more than one-third of the world's population. Other states such as Kiribati have fewer residents than an English county.

One state in the early twenty-first century is able to project its military power globally and qualifies as a super-power (the United States). A small number of states have military forces sufficiently capable of exercising power regionally, for example, Japan, China, Britain, France, and Russia. In addition, other states than these have also developed nuclear weapons which make them credible threats in regional arenas, for instance North Korea, India, Pakistan, and possibly Iran. Most states in sub-Saharan Africa and many Asian, Middle Eastern, and Latin American states lack the military means to defend national interests.

Nation-states also vary enormously in their degree of cultural and social integration. A small number are the heirs (or joint legates) to great and long-lasting cultures, such as the Chinese, Japanese, Indian, Islamic, Western, and Russian Orthodox civilizations. Yet the fault lines across these civilizations often have produced international conflict. Most states, however, contain more than one "nation" in the sense of a community of persons sharing values and envisioning a common future. They may be divided by race, ethnicity, language, or religion, and both cultural and social divisions inhibit the nation's ability to address common problems. Nigeria, for example, struggles to establish a unified national identity among its various tribal, regional and religious groups.

An extremely conspicuous difference among nation-states is their level of economic development. A minority of the world's states are economically developed countries (EDCs) or rich nations. One standard for economic development is a per capita Gross Domestic Product (GDP) of $10,000, and by this definition over 40 states have entered the rich nations' club. With few exceptions, these nations are located in the northern hemisphere; and often for this reason collectively they are referred to as the "North." The other nations mostly are located in equatorial zones or in the southern hemisphere (the South); typically they are loosely labeled "lesser developed countries" (LDCs) or just "developing countries." The label is inexact as the economies of some are developing quite rapidly (for example, the "newly-industrialized countries" (NICs), including South Korea, Taiwan, Singapore, Hong Kong, Malaysia, Thailand, and Brazil, some of which are now considered economically developed). A much larger number of countries lie at the bottom of the heap; they are considered the "poorest of the poor," because their per capita GDP is less than $400 or one dollar a day. Some 1.1 billion people live in such countries.

We pay attention to the economic development levels of countries because often they help us explain cross-national differences in environmental policies and outcomes. The rich countries, which in the past and presently have experienced rapid industrial development, have also disproportionately exploited the world's resources and created much of the global environmental problems. Yet they have the economic means to mitigate pollution of land, air, and water, and to adapt to rapid environmental change. Poor countries, on the other hand, are less culpable for the emerging global environmental crises such as climate warming. They are also less able, because of limited economic development, to address problems of pollution and adapt to environmental change.

Although the focus of this study is the nation-state, we consider at length one super-state, the European Union (EU). In 2006, the EU contains 25 member states. EU member states have ceded some of their sovereignty in the environmental policy area, but they remain solely responsible for implementation. In general, the EU resembles a political system that is "multilevel, horizontally complex (and) evolving."[16]

4. A POLITICAL APPROACH

Table 1.1 lists the nation-states used as examples in various parts of this book. Following the World Bank classification scheme they are divided, economically, by per capita national income. They are further categorized by regime type and core characteristics of their political and economic systems. The dominant type of interest group representation is noted as a rough indicator of state-society relations. These identifiers are typical of comparative

Table 1.1. Selected nation-states – political and economic characteristics

Nation-state	Regime type*	System type†	Territorial distribution‡	Interest groups§	ESI score/rank**
High income (US$10,066 and above per capita national income in 2004)					
Australia	Democratic	Parliamentary	Federal	Pluralist	61.0/13
Brunei	Authoritarian	n/a	Unitary	n/a	n/a
Canada	Democratic	Parliamentary	Federal	Pluralist	64.4/6
France	Democratic	Semi-pres.	Unitary	Mixed	55.2/36
Germany	Democratic	Parliamentary	Federal	Corporatist	56.9/31
Gr. Britain	Democratic	Parliamentary	Unitary	Mixed	50.2/65
Greece	Democratic	Parliamentary	Unitary	Corporatist	50.1/67
Italy	Democratic	Parliamentary	Unitary	Corporatist	50.1/69
Japan	Democratic	Parliamentary	Unitary	Mixed	57.3/30
Norway	Democratic	Parliamentary	Unitary	Corporatist	73.4/2
Monaco	Authoritarian	Parliamentary	Unitary	n/a	n/a
Netherlands	Democratic	Parliamentary	Unitary	Corporatist	53.7/40
Singapore	Authoritarian	Parliamentary	Unitary	n/a	41.84/na
S. Korea	Democratic	Presidential	Unitary	Mixed	43.0/122
Spain	Democratic	Parliamentary	Federal	Corporatist	48.8/76
Sweden	Democratic	Parliamentary	Unitary	Corporatist	71.7/4
Taiwan	Democratic	Semi-pres.	Unitary	Mixed	32.7/145
U.S.	Democratic	Presidential	Federal	Pluralist	52.9/45

* "Transitional" refers to democratizing regimes where the outcomes remain uncertain. Otherwise we use a simple dichotomy of democratic and authoritarian, the latter including civilian, military, monarchical and theocratic regimes.

† "Semi-pres." (semi-presidential) systems elect parliaments and powerful heads of state.

‡ Refers to the distribution of power between national government and the geographical subdivisions of the nation-state (e.g. states or provinces).

§ "Pluralist" refers to states with few formal institutional relationships between interest groups and state agencies or political parties. "Corporatist" refers to all types of formal linkages between organized groups and state agencies or political parties, including "neo-corporatist" and "party corporatist." "Mixed" connotes significant formal linkages between states and/or parties for some groups and issue areas, and the existence of influential unaffiliated groups.

** "The Environmental Sustainability Index (ESI) benchmarks the ability of nations to protect the environment over the next several decades. It does so by integrating 76 data sets—tracking natural resource endowments, past and present pollution levels, environmental management efforts, and the capacity of a society to improve its environmental performance—into 21 indicators of environmental sustainability. These indicators permit comparison across a range of issues that fall into the following broad categories:

- Environmental Systems
- Reducing Environmental Stresses
- Reducing Human Vulnerability to Environmental Stresses
- Societal and Institutional Capacity to Respond to Environmental Challenges
- Global Stewardship"

[Yale Center for Environmental Law and Policy, Yale University, and Center for International Earth Science Information Network, Columbia University. *2005 Environmental Sustainability Index: Benchmarking Environmental Stewardship*. Available at www.yale.edu/esi, 1.]

Nation-state	Regime type	System type	Territorial distribution	Interest groups	ESI score/rank
Upper-middle income (US$3,256-10,065 per capita income in 2004)					
Barbados	Democratic	Parliamentary	Unitary	n/a	n/a
Czech Rep.	Democratic	Parliamentary	Unitary	n/a	46.6/92
Dominica	Democratic	Parliamentary	Unitary	Mixed	n/a
Grenada	Democratic	Parliamentary	Unitary	Pluralist	n/a
Hungary	Democratic	Parliamentary	Unitary	n/a	52.0/54
Malaysia	Transitional	Parliamentary	Federal	Mixed	54.0/38
Mexico	Transitional	Presidential	Federal	Corporatist	46.2/95
Palau	Democratic	Presidential	Unitary	n/a	n/a
Poland	Democratic	Parliamentary	Unitary	n/a	45.0/102
Russia	Transitional	Semi-pres.	Federal	Mixed	56.1/33
Saint Lucia	Democratic	Parliamentary	Unitary	Pluralist	n/a
Slovak Rep.	Democratic	Parliamentary	Unitary	n/a	52.8/48
Trinidad & Tobago	Democratic	Parliamentary	Unitary	n/a	36.3/139
Venezuela	Democratic	Presidential	Federal	Pluralist	48.1/82
Lower-middle income (US$826-3,255 per capita national income in 2004)					
Brazil	Democratic	Presidential	Federal	Pluralist	62.2/11
China	Authoritarian	Party-state	Unitary	n/a	38.6/133
Cuba	Authoritarian	Party-state	Unitary	n/a	52.3/53
Egypt	Authoritarian	Presidential	Unitary	n/a	44.0/115
Indonesia	Transitional	Presidential	Unitary	n/a	48.8/75
Iran	Authoritarian	Semi-pres.	Unitary	Mixed	39.8/132
Kiribati	Democratic	Semi-pres.	Unitary	n/a	n/a
Thailand	Democratic	Parliamentary	Unitary	n/a	49.7/73
Ukraine	Transitional	Semi-pres.	Unitary	n/a	44.7/108
Vanuatu	Democratic	Parliamentary	Unitary	n/a	n/a
Low income (US$825 or less per capita national income in 2004)					
India	Democratic	Parliamentary	Federal	Pluralist	45.2/101
Kenya	Transitional	Presidential	Unitary	n/a	45.3/100
Nigeria	Transitional	Presidential	Federal	Pluralist	45.5/98
N. Korea	Authoritarian	Presidential	Unitary	n/a	29.2/146
Pakistan	Authoritarian	Semi-pres.	Federal	n/a	39.9/131
Viet Nam	Authoritarian	Party-state	Unitary	n/a	42.3/127

political studies that emphasize institutions and state-society relations and are not selected with special reference to environmental political issues.[17] "High income" nation-states are mainly industrial and post-industrial democracies. But important exceptions include the oil rich, low population countries of the Middle East such as Saudi Arabia and Kuwait, and newly industrialized countries that have achieved high-income status, such as the so-called East Asian "tigers," South Korea, Singapore, Hong Kong and Taiwan. "Upper-middle" and "lower-middle" income countries are mixed groups, including a variety of industrializing and resource dependent economies, as well as several consolidated democracies, transitional and authoritarian regimes. Of particular interest among these groups are newly industrializing countries such as Brazil and Mexico, and transitional countries such as Russia and the Ukraine since industrialization and rapid economic change are frequently associated with environmental stresses. "Low income" countries include a variety of poor, resource dependent and/or agricultural economies which are politically weak and often struggling with the legacies of authoritarianism, ethnic conflict and political violence. But there are also notable exceptions that warrant special attention such as India—a consolidated democracy with a rapidly modernizing economy—and Nigeria—an oil-rich state with a large population currently struggling to establish democratic rule. The table is not meant to be exhaustive, and other characteristics will be considered along the way. The variations listed in Table 1.1, along with differences in national territory and population, military power, and culture, will figure in our discussion of environmental problems, as they may have the power to explain differences in national approaches to environmental issues. Our subject throughout, however, is the political system. The thesis of this book is that political system characteristics may explain cross-national differences in environmental policy as satisfactorily as any of the variations among nations discussed above.

Five questions organize the study:

1. Which actors make decisions on environmental issues, informed by what values?
2. What national processes connect actors, institutions, and the people?
3. Which institutions, in what configurations, are responsible for resolving environmental problems?
4. What effect do differences in actors, institutions and processes have on environmental outcomes domestically?
5. What effect do these differences in political system characteristics have on the responses of nations to global environmental problems?

These questions organize our approach to the understanding of differences in environmental outcomes of the world's nations.

5. RESEARCH IN COMPARATIVE ENVIRONMENTAL POLITICS

As a systematic field of study, comparative environmental politics is relatively new, and the research literature is limited. Most of what we know about different countries' environmental policies and outcomes is based on single-country descriptive studies. Such studies are available for all the developed nations, for example, Rosenbaum's *Environmental Politics and Policy*[18] and Broadbent's *Environmental Politics in Japan.*[19] Coverage of environmental politics of developing nations is more limited. Nevertheless, there is a growing literature on the larger and more powerful ones, such as China, as noted in the recent publication of Judith Shaprio's *Mao's War Against Nature*[20] and Elizabeth Economy's *The River Runs Black.*[21]

Studies comparing environmental politics and outcomes in two or more nations are less abundant. Among the earliest of this type of study was David Vogel's comparison of the United States and Great Britain, *National Styles of Regulation*, published in 1986.[22] Yet within the last decade, a number of books and monographs have examined more than two countries. Some of these "small N" studies survey countries at comparable levels of economic development. For example, Desai's *Ecological Policy and Politics in Developing Countries* (1998) features chapters on China, Taiwan, India, Venezuela, Nigeria, Indonesia, Mexico, Thailand, and Czech/Slovakia.[23] Although each chapter is written by a different author (s), they follow common themes. A smaller number of countries, in one large world region, is treated by Liu and So's *Asia's Environmental Movements.*[24] The focus of the chapters in this volume is on the impact of development level on environmental policy. Another regional comparative approach is found in the study by Weale et al., of *Environmental Governance in Europe*. In addition to treatment of governance issues in the EU as a whole, the volume examines in some detail the national environmental policies and politics of six member states: Germany, Greece, Italy, Spain, the Netherlands, and the United Kingdom.[25]

The latest edited work by Desai, *Environmental Politics and Policy in Industrialized Countries* (2002)[26] continues this tradition with respect to EDCs, treating the U.S., Britain, Germany, Japan, Canada, Italy, and Australia. Again Desai asks chapter authors to treat three elements: description of environmental policies and problems; the environmental policy process; and the effectiveness of environmental policies and regulations in dealing with environmental problems. The presentations are both descriptive and qualitative. Perhaps the most effective recent small N study of environmental politics in EDCs is Miranda Schreurs' *Environmental Politics in Japan, Germany, and the United States.*[27] She asks how three highly developed nation-states—the U.S., Germany, and Japan—came to have such different policies toward global environmental problems such as climate change, and points to political processes and institutions which appear to explain the different outcomes. A related approach is adopted by Dryzek et al., who emphasize environmental parties and movements in the United States, UK, Germany, and Norway.[28]

A few qualitative studies compare the responses of countries at different stages of economic development to large problems in environmental policy and administration, for instance implementation deficits. Janicke and Weidner, editors of *National Environmental Policies: A Comparative Study of Capacity-Building* (1997), attempt to ascertain the reasons explaining why some nations became environmental "pioneers" or "models," while others have fallen behind and become "laggards," or in general are incapacitated and unable to implement effective environmental policies. A similar approach to a different topic is the special issue of *Environmental Politics* on "Green Parties in National Governments" (2002).[29] And Bron Taylor's very readable compilation of environmental movement cases from throughout the world—*Ecological Resistance Movements*—follows this pattern by sampling from both EDCs and LDCs.[30]

Data restrictions, and in particular the absence of systematic and comprehensive information on national emissions of pollutants or biodiversity losses, deforestation, and desertification, have frustrated the development of "large N" quantitative comparative studies of environmental politics and policy. Nevertheless, the journal literature has several path-breaking studies of comparative differences among EDCs. For example, Crepaz studies the impact of corporatism on national variation in air pollution levels of western nations,[31] and Scruggs examines the relationship between institutions and environmental performance in 17 western democracies.[32]

Although the professional literature in comparative environmental politics is not vast, it is sufficient to generate interesting hypotheses and examine generalizations. Writing this book would not have been possible without the contributions made by hundreds of researchers, narrating the experiences of countries with quite different political, economic, and social systems, from throughout the world.

6. PLAN OF THE BOOK

This study unfolds in five substantive chapters. Chapter 2 considers four aspects of state-society relationships. First, it considers traditional attitudes and values toward the environment, based on an understanding of political culture—the beliefs, attitudes, and opinions in the minds of citizens as they contemplate their environment. Specifically, we discuss the beliefs citizens have about relationships between humans and the environment in select countries (for example, whether they regard the environment as possessing intrinsic value), and the attitudes of citizens and elites toward environmental policy. Second, we explore the relationship between economic development and social change, with a particular focus on modernization, political development, and dependency theories developed by scholars to explain these changes. Third, we treat the development of a "new environmental paradigm," focusing on sustainable development, and see how this is reflected in public opinion cross-nationally. Fourth, we consider the different types of state-society organization, centering the discussion on pluralism versus corporatism, and then examine the impact this has on environmental decision-making.

In chapters 3, 4, 5 and 6 we examine the practical politics of the environment; actively comparing interests, institutions, and policy processes in developed democracies, transitional and developing states (including cases from both underdeveloped and newly industrialized states). Chapter 3 asks large questions about the diversity of groups, organizations, and movements throughout the world pursuing improvement of the environment, which pull together the ideas, attitudes, and values identified in the previous chapter and connect them to the political institutions discussed in chapter 4. The chapter begins by investigating environmental interest groups or non-governmental organizations, often called ENGOs. Then it treats environmental movements, both in EDCs and LDCs, with a special treatment of ecological resistance movements. The core of the chapter is analysis of the role that organized environmental perspectives play—as political parties—mostly in western nation-states, but with some comparative references to the developing world as well. It asks what conditions lead to development and growth of Green parties (including the role of electoral institutions), to their participation and fortunes in government, and to the propensity for established or mainstream political parties to adopt environmental perspectives as part of their electoral and governing strategies. The chapter briefly considers the important role of media in disseminating environmental perspectives and news about environmental organizations. It concludes with a discussion of the relationship between the process of democratization and political mobilization on environmental issues.

Chapter 4 introduces the structure and organization of the state itself, and the role that its political institutions and arrangement of authorities and powers play in the development of environmental policy. After describing the nature of institutions, the chapter asks what difference constitutional limitations on state power and the formation of liberal state systems makes in environmental policy-making. Second, the chapter considers the geographic distribution of authority between the national government and lower levels, such as states, provinces, or municipalities, and its bearing on environmental policy outcomes. This section compares federal, confederal, and unitary state systems, but it also considers the actual amount of decentralization within systems. The third feature of the institutional environment is whether powers are separated among different branches of government—as in the U.S.

presidential system—or concentrated, as one sees in most parliamentary systems. It reviews corporatist systems as a special subset of parliamentary systems, and then asks whether courts have the independence to limit coercive government action and/or mandate environmental outcomes. This chapter concludes with an analysis of the political opportunity structure of nations with respect to environmental agendas.

Chapter 5 examines the concept of political capacity to effect environmental policy in a nation, and asks who is likely to exercise that capacity. It then considers how the resources of the state are arranged with respect to implementing environmental policy, considering economic, human, and political resources. Next the chapter discusses administrative competence in environmental policy-making; it compares and contrasts national environmental institutions and strategic environmental planning. The chapter compares the environmental capacity of nation-states, identifying the pioneers, models, and laggards, and the relative differences in "implementation deficits." This chapter also explores issues of globalization, international networks and linkages, and the roles of international organizations and international ENGOs in augmenting the political capacity of developing states.

Chapter 6 introduces two global environmental problems: climate change and biodiversity loss. In each area, it considers the scientific nature of the issue, international action to resolve the problem, and responses of both EDCs and LDCs to the problem. The chapter asks whether we can differentiate "national styles" in responses of countries to global environmental crises.

Chapter 7 concludes the volume, by returning to the questions introduced in Chapter 1. It illustrates the difficulties in defining, adopting, and implementing sustainability policies, and attempts to assess the different roles of economic and political variables in explaining environmental change.

---

[1] http://news.bbc.co.uk/1/hi/world/south_asia/4064527.stm.
[2] See, for example, the Exxon Valdez Oil Spill Trusteeship Council, http://www.evostc.state.ak.us/History/index.htm.
[3] News Online, http://news.bbc.co.uk/1/hi/world/americas/101035.stm; and http://news.bbc.co.uk/1/hi/world/americas/476540.stm.
[4] "World 'failed' Bhopal gas victims," http://news.bbc.co.uk/1/hi/world/south_asia/4050739.stm.
[5] Exxon Valdez Oil Spill Trusteeship Council, "Five Years Later: Status Report on the Exxon Valdez Oil Spill," Anchorage: The Oil Spill Public Information Center, 1994.
[6] Green Nature, "Air Pollution in Mexico," http://greennature.com/article827.html.
[7] The World Bank, "Improving Air Quality in Mexico City," http://web.worldbank.org/WBSITE/EXTERNAL/COUNTRIES/LACEXT/MEXICOEXTN/0,,contentMDK:20047307~menuPK:338419~pagePK:141137~piPK:141127~theSitePK:338397,00.html.
[8] See Mattei Dogan and Dominique Pelassy. *How to Compare Nations: Strategies in Comparative Politics*, 2nd edition. Chatham, NJ: Chatham House Publishers, Inc., 1984.

[9] See David Apter. "Comparative Politics, Old and New." In Robert E. Goodin and Hans-Dieter Klingemann, editors. *A New Handbook of Political Science.* Oxford: Oxford University Pres, 1996.
[10] See, for example, David Held, *Global Covenant: the Social Democratic Alternative to the Washington Consensus*. Cambridge: Polity Press, 2004, 73-88.
[11] James N. Rosenau, "Turbulence in World Politics: A Theory of Change and Continuity," in *Global Politics and a Changing World: A Reader*, edited by Richard W. Mansbach and Edward Rhodes. Boston and New York: Houghton Mifflin, Co., 2003, 21.
[12] Steven D. Krasner, "Sovereignty," *Foreign Policy* (January-February 2001), 2.
[13] The classic statement of this debate remains, Kenneth N. Waltz, *Man, the State, and War: A Theoretical Analysis*. New York: Columbia University Press, 1959.
[14] Elizabeth R. deSombre, for example, makes the domestic politics argument in explaining U.S. reluctance to ratify international environmental treaties in "Understanding United States Unilateralism: Domestic Sources of U.S. International Environmental Policy, in Regina S. Axelrod, David L. Downie, and Norman J. Vig, eds. *The Global Environment: Institutions, Law, and Policy*. Washington DC: CQ Press, 2005, 181-199.
[15] Norman J. Vig, "Introduction: Governing the International Environment," in Axelrod, Downie and Vig. 13-14.
[16] See Albert Weale, Geoffrey Pridham, Michelle Cini, Dimitrios Konstadakopulos, Martin Porter, and Brendan Flynn, *Environmental Governance in Europe*. New York: Oxford University Press, 2000, 1.
[17] Some key characteristics, such as type of electoral system, are not included in the table because the large number of variations and exceptions makes them difficult to display effectively. Nevertheless, the effects of electoral systems are essential to comparative political analysis and will be examined in particular cases, as in the comparison of the United States and Germany in this chapter.
[18] Walter, A. Rosenbaum, *Environmental Politics and Policy*, 5th ed. Washington, DC: Congressional Quarterly Press, 2002.
[19] J. Broadbent, *Environmental Politics in Japan: Networks of Power and Protest.* Cambridge: Cambridge University Press, 1998.
[20] Judith Shapiro, *Mao's War Against Nature: Politics and the Environment in Revolutionary China*. Cambridge: Cambridge University Press, 2001.
[21] Elizabeth C. Economy, *The River Runs Black: The Environmental Challenge to China's Future.* Ithaca, NY: Cornell University Press, 2004.
[22] David Vogel, *National Styles of Regulation: Environmental Policy in Great Britain and the United States*. Ithaca, NY: Cornell University Press, 1986.
[23] Uday Desai. *Ecological Policy and Politics in Developing Countries.* Albany, NY: State University of New York Press, 1998.
[24] Yok-shiu F. Lee and Alvin Y. So, editors. *Asia's Environmental Movements: Comparative Perspectives.* Armonk, NY: M. E. Sharpe, 1999.
[25] Weale et al., 2000.
[26] Uday Desai. *Environmental Politics and Politics in Industrialized Countries.* Cambridge, MA: The MIT Press, 2002.
[27] Miranda A. Schreurs. *Environmental Politics in Japan, Germany, and the United States.* Cambridge: Cambridge University Press, 2002.
[28] John S. Dryzek, D. Downes, C. Hunold, D. Schlosberg, with H-K. Hernes, *Green States and Social Movements.* Oxford: Oxford University Press, 2003.
[29] Ferdinand Muller-Rommell and Thomas Poguntke, guest editors. "Green Parties in National Governments," special issue of *Environmental Politics*. Vol. 11, No. 1 (Spring 2002). This journal remains the premier source for articles on comparative environmental politics.

---

[30] Bron R. Taylor, editor. *Ecological Resistance Movements: The Global Emergence of Radical and Popular Environmentalism.* Albany, NY: State University of New York, 1995.
[31] See *Environmental Politics,* Vol. 9, No. 3 (Fall 1999).
[32] Lyle A. Scruggs. "Institutions and Environmental Performance in Seventeen Western Democracies." *British Journal of Political Science*, Vol. 29 (1999), 1-31.

CHAPTER 2. STATE-SOCIETY RELATIONS

1. CONCEPTUAL DISTINCTIONS

"State" and "society" are concepts at the heart of the social sciences. As abstractions they denote, in the case of the state, the political system (to most, the "government"), the organized and integrated system of institutions, relationships, and rules able to imperatively command behavior. The "society," on the other hand, represents virtually everything else—the marketplace, religious organizations and communities, ethnic groups (families, tribes, clans), voluntary associations (from civic and social service organizations to sports, recreation and cultural organizations), social relationships, and spontaneously arising social movements.

Social contract theorists, beginning with the publication of Thomas Hobbes' *Leviathan* in 1651 during the playing out of the English Civil War and Locke's writing of the *Two Treatises of Civil Government* in 1688, were the first to conceive of "society" as potentially free-standing and a counterpoise to the state. Their conception of an autonomous society was critical to the development of governments based on individual consent and the rise of democratic thought in the West. The model of an autonomous society provided a safe haven for the individual oppressed by the state, and thus seemed a limit to arbitrary state power. It is from this concept that some western societies derive their principles of constitutionalism, limited government, individual rights and liberties.

The state-society dichotomy, thus conceived, has played an important part in debates over environmental policy. For example, U.S. political culture tends to resist extensive government regulation of the environmental impacts of private actions (economic and otherwise). Reflecting a libertarian tradition derived from Locke, in debates over environmental policy, U.S. policy-makers often dichotomize these differences as free-market or voluntary solutions versus "command-and-control" or coercive solutions. Many Western European and developing states, basing state-society relations on social-democratic or Marxian traditions, see the state as both a guarantor and protector of human interests. These political cultures, many of which are also democratic, tend to be more amenable to regulatory policies as a means of providing for the public or general good. The assumption of the former is that markets self-regulate, distributing values of all kinds more efficiently and equitably than any political agency. The assumption of the latter is that the social forces embedded in free markets do not always distribute the costs and benefits of economic activities equitably but can be corrected through democratic political processes.

These distinctions and relationships between state and society are of particular importance for understanding environmental politics. Throughout the world, environmental politics has tended to originate as societal reactions to the negative and presumably unintended effects of economic activities (i.e., negative externalities) and government policies (ranging from fiscal and regulatory policies, to development projects, to war). Whether arising as "new social movements" or taking the form of conventional interest group or political party activities, environmentalism has become an important component of an independent and voluntary realm of political activity—what democratic theorists refer to as "civil society." States have also reacted to external pressures from other states, international organizations and international environmental non-governmental organizations (ENGOs—see chapter 3) by creating new environmental protection institutions and policies. In many cases international ENGOs become influential not only by pressuring governments directly, but through the establishment of information and support networks that raise awareness of environmental issues and provide political resources for domestic civil society organizations in their own attempts to influence national policy.[1] Additionally, environmentalism has generated or revived interest in forms of democratic decision making that take place outside of the formal organs of the state authority such as deliberative, participatory and direct democracy at the community level. Theorists focusing on these latter developments have argued that environmentalism may both revitalize established, liberal democracies (as in the United States or Western Europe) *and* contribute to the weakening of authoritarian regimes (as in Mexico and the Soviet Union).[2]

We will be comparing the patterns of state-society relations cross-nationally, based on the available published literature, in the belief that these patterns offer insight into nation-state responses to domestic and global environmental events. This area of investigation also is historically contingent, and it stresses path dependency—that is, decisions made in the past impede or enhance the likelihood of similar decisions (or decision processes) in the present.[3]

This chapter considers four aspects of state-society relationships: traditional attitudes and values toward the environment; economic development and social change; development of a "new" environmental paradigm; and social organization and state decision-making.

2. TRADITIONAL ATTITUDES AND VALUES TOWARD THE ENVIRONMENT

Comparative political analysis has undergone several paradigm shifts over time, and several paradigms coexist. Originally, the sub-discipline was a formal enterprise of comparing national laws and statutes, constitutions and codified political structures and processes. In the post World War II era, however, comparative political studies discovered the importance of culture, society, the non-western world, and the need to explain political behavior. The unhappy experiences with fascist totalitarianism before and during the war sparked an interest in understanding where, why and under what conditions nation-states became democratic or dictatorial. Informed by anthropologists such as Ruth Benedict, political scientists pursued "national character studies," attempting to understand the modal personalities that lead to autocracy, democracy, and social volatility. The cold war contests between liberal-capitalist and state-socialist systems and the massive wave of decolonization that followed the war—beginning in India in 1947 and engulfing most of Africa, Southeast Asia and the Caribbean by the 1950s-70s—intensified the interests of scholars and policy makers in political development. Since most states, including the Soviet Union, had well crafted constitutions, elaborate electoral systems and representative bodies, by the 1950s it had become clear that formal documents and procedures, rather than determining the politics of a country, were themselves determined by forces outside of the strictly circumscribed realm of government and the state. While these original attempts to bring culture and behavior into the picture proved problematic, they eventually led to more rigorous, nuanced and less ethnocentric studies—including well-articulated theories of political culture, political systems, and structural-functionalism—under the general rubric of behavioralism.[4] Therefore, cultural influences are our starting point, although as we present cases and examples we will modify the often overly deterministic nature of behavioralist analysis by considering the conditioning effects of economic factors, social structure, and political institutions as well as the growing influence of globalization.

Attitudes and values (as well as beliefs and opinions) are patterns in the mind of citizens as they contemplate their universe and interact with others. We assume that these psychological states translate into action: that values and attitudes influence behavior. For more than a generation, social scientists have studied this relationship and found that, generally, culture and values tend to predict behavior, when unconstrained by exogenous factors (for instance, threats to one's survival) and more powerful internal drives (such as hunger). A second assumption in comparative politics research is that values and attitudes vary across cultural regions and may vary from country to country (and within diverse countries too).

In their path-breaking studies of politics in developing areas after World War II, political scientists Gabriel Almond and Sydney Verba popularized the concept "political culture" as a tool for the systematic study of differences in values and attitudes cross-nationally. By political culture, they meant citizens' orientations toward the political system, the political and policy-making process, and policy outputs and outcomes.[5] Our study of comparative environmental politics focuses on the first and third levels: the beliefs citizens have about relationships between humans and the environment (whether, for example, they regard the environment as of intrinsic value), and the attitudes of citizens and elites toward environmental policies.

We will examine a few traditional sets of values and attitudes toward the environment, beginning in the West. In the process, we will be making generalizations that should be viewed critically, because the tools for understanding values and attitudes within countries, as well as cross-nationally, are not well-developed. And while it is difficult to prove causal relationships between values and practices, we offer some historical examples that correlate attitudes and values about nature with practices that have left lasting marks on the environment.

2.1. Western Values and the "New World"

The values of early American settlers toward the environment emphasized freedom and individuality. People then tended to believe that the land should be used as they wished. Arriving in an apparently untamed wilderness (forgetting previous occupation by American Indians), white settlers believed in their ability to dominate. As they developed improved agricultural technology and then industrial machinery, they believed that it should be used to master nature. Also, from their perspective, there were no limits to growth. Lester Milbrath calls this the "dominant social paradigm," and dates it to the founding of America by settlers from Europe.[6]

John Muir describes this attitude of hostility to nature in his description of a farmer:

> Not content with the so-called subjugation of every terrestrial bog, rock and moorland, he would fain discover some method of reclamation applicable to the oceans and the sky, that in due calendar time they might be bought to bud and blossom as the rose . . . Wildness charms not my friend.[7]

This traditional view treats all of nature as an object for human use. Throughout the nineteenth and early twentieth centuries, then, the U.S. government supported and encouraged economic development aimed at "improving" the land by exploiting natural resources for economic growth and enhancing national political and military strength. Government programs subsidized the infrastructure needed to support modern forms of agriculture, resource extraction, industry and commerce, and helped establish new settlements to accommodate an expanding population and integrate new lands into the national economic, political and social systems. Policies supporting this social paradigm included wars against and forced resettlement of indigenous peoples, the Homestead Act of 1862, and large grants of public lands to private entrepreneurs as incentives for building railroads. Diversions from this strategy, when they occurred, were not aimed at basic cultural or ideological changes, but rather at either conserving valuable natural resources from utter depletion or preserving charismatic landscapes and creatures from destruction. Conservation strategies might include the replacement of logged-out forests with plantations of rapidly growing species to insure steady supplies of wood products. Preservation strategies included the creation of the first national parks.[8]

The southern part of the western hemisphere was treated similarly by its European settlers. Although the patterns, methods, and many of the original motivations of conquest and settlement were different, the Spanish and Portuguese colonial enterprises in the Americas displayed similar attitudes toward the natural environment. In explaining Spain's drive to establish an empire in the new world, the old grade-school aphorism, "gold, God and glory," holds up surprisingly well. Within two years of Columbus' first voyage, Pope Alexander VI, in the Treaty of Tordesillas, divided the largely unexplored continent between the two great Catholic monarchies of the day—Spain and Portugal. The royal houses were charged to take dominion over the land and people they found, extracting riches for themselves and the Church, and collecting the souls for the Church. The cornerstones of Spanish colonial rule, the *encomienda* and the royal land grant, gave conquistadors control over the land and the labor of the people on the land. Accompanied by Catholic clergy who took responsibility for the conversion and protection of the indigenous populations, the conquistadors were expected to find, extract and export new sources of material wealth (primarily precious metals) for the benefit of the Crown. Failing that, the land would be developed for agriculture to feed growing European tastes for sugar and other tropical products, provide dyes for European textiles,[9] and feed the growing numbers of miners, artisans, slaves and colonial administrators. All of this involved the clearing of forests, the growth of towns and cities, and the settlement of semi-nomadic indigenous populations in agricultural and mining communities.[10] The effects of colonial rule and plantation agriculture on Cuban forests illustrate European beliefs in human dominion over nature that have permanently affected the landscape.

When Christopher Columbus arrived in Cuba, approximately 9 million hectares (more than 90 percent of its surface area) were forested. As of 1995, the Cuban government estimates that 1.5 million hectares of dense forestland are left.[11] Initially, Cuban forest products were prized and protected by Spanish imperial authorities. Accounts of Columbus' voyages include rapturous descriptions of the dense and fragrant forests and the marvelous properties of their hardwoods. Cuban forests were protected under Spanish colonial law so that Cuban wood products could service the Crown as construction materials for government and church buildings in Spain and to build warships to protect the empire. It was not until the eighteenth century with the introduction of large-scale sugar production that the Cuban forests were overexploited. As the economy and power structure of colonial Cuba changed, so did its forestry laws. In 1815 sugar planters and mill owners succeeded in getting the Spanish parliament to lift most restrictions on the clearing of woodlands. According to one estimate, by 1830 the sugar industry accounted for deforestation at an annual rate of 26,800 hectares—half for fuel, half for clearing land for planting and milling.[12] This attitude toward nature largely survived independence from Spain in 1898 and the socialist Revolution of 1959. Only after the fall of the Soviet Union, and the end of a guaranteed market for its sugar, has the Cuban government seriously attempted to diversify away from sugar production and promulgate effective forest conservation policies.

2.2. Asian Values

In traditional China, there were several perspectives on the relationship of humans and nature. The tradition of Taoism saw humans as one with nature. As the Taoist sage Lao Tzu remarked:

> The universe is sacred.
> You cannot improve it.
> If you try to change it, you will ruin it.
> If you try to hold it, you will lose it.[13]

Thus, humans should accommodate nature and not attempt to transform it. A second Chinese tradition, Buddhism, expressed an ethic of reverence toward all living things. Each living entity had a divine essence, which commanded respect. However, Confucianism became the dominant social ethic. It was an humanitarian ethic, focusing on norms of human behavior. While humans ought to live in harmony with all life and their ecosystem, attitudes toward nature were instrumentalist. They encouraged exploitation and management of nature for human purposes. Shapiro comments that Confucianism is "an anthropocentric ethic that espouses harmony with nature but understands nature primarily as a resource for human beings, to be shaped to human desires, though in accordance with its own laws."[14]

As Confucianism and then neo-Confucianism penetrated into East and Southeast Asian regions influenced by Chinese culture, the emphasis on humanity (also translated as "benevolence") moderated. For example, in Japan, the increasingly dominant view became one of unity of humans and nature, a view which became part of the popular culture through works such as the *Precepts for Children* of Kaibara Ekken:

> (N)ot only do all men at the outset come into being because of nature's law of life, but from birth till the end of life they are kept in existence by the support of heaven and earth. Man surpasses all other created things in his indebtedness to the limitless bounty of nature. It will be seen therefore that man's duty is not only to do his best to serve his parents . . . but also to serve nature throughout his life in order to repay his immense debt.[15]

When we turn to the traditional values and attitudes of South Asia and specifically India, we find an equally broad range of diversity, including systems and subsystems of Hinduism, several schools of Buddhism and Jainism, as well as the materialism of the Carvaka. In general, however, the emphasis of values is on the spiritual and not the material. With the exception of the Carvaka, humans and nature are not regarded as physical in essence; material welfare is not a goal of human life. Indeed, as the *Upanishads* comment: "One should know that Nature is illusion."[16] This spiritual element permeates the more recent but still traditional Gandhian environmental ethic of voluntary simplicity and self-restraint. One of Gandhi's best known aphorisms expresses this: "The world has enough for everybody's need, but not enough for everybody's greed."[17]

2.3. Islamic and African Values

In the Middle East, Africa, and Asia, environmental values and attitudes have been influenced strongly by Islam. In the Islamic view, the universe is Allah's creation, and it cannot operate without control and guidance from the Almighty. To protect life on earth, humans are expected to discharge their responsibility to Allah and the environment.[18]

Islam makes kindness to animals a part of its faith, as the following passages represent in the teachings of Prophet Mohammed:

> The Prophet was asked if acts of charity, even to the animals, were rewarded by God. He replied: 'Yes, there is a reward for acts of charity to every beast alive.' The Prophet told his companions of a serf, who was blessed by Allah, for saving

> the life of a dog by giving it water to drink, thereby quenching its thirst. We were on a journey with the Apostle of God, and he left us for awhile. During his absence, we saw a bird called hummara with its two young and took the young ones. The mother bird was circling above us in the air, beating its wings in grief, when the Prophet came back and said: 'Who has hurt the feelings of this bird by taking its young? Return them to her.'[19]

In Sub-Saharan Africa there are a variety of cultural traditions identifying the place of human beings in nature, but as Ben Wisener cautions, Western social science is likely to be inadequate for a full understanding of African attitudes toward the natural world. Nevertheless, a few key aspects can be identified. Instead of being conceptually separate, spirituality, human survival in the temporal world, and ecological values and principles are fully integrated. What western observers might construe as attitudes toward the environment in contemporary Africa are actually much broader, and substantially different than simple environmentalism. Present-day African attitudes toward the environment reflect both struggles for preservation of nature and to maintain stable livelihoods as well as efforts to retain or revive deeply held beliefs about the relationships of humans and all living things to the earth. According to Wisener, understanding human attitudes toward the natural environment in Africa requires an understanding of the "lifeworlds" of its indigenous peoples—a concept that is more inclusive and fluid than environment or ecosystem. Traditional forms of agriculture, the taking of plants and animals for food, medicine and ritual, the interment of family members and the continual presence of their spirits in the community, and the cultural and physical survival of tribal groups are of one piece. Contrary to dominant Western notions of value, the valuation of land and resources emphasizes the spiritual and social rather than the economic. The relationship with the land, its resources, fauna and flora is identical to the meaning, integrity and survival of the human communities that are a part of it.[20]

The relationship of African peoples to the land is by no means static or homogenous, however. Traditionally, some lands have been considered sacred, and much land "ownership" has been communal or clan-based. But post-independence governments have frequently continued the western-style private property and development regimes inherited from colonial days. The *kaya* forests of coastal Kenya, for example, are considered sacred by the Mijikenda peoples and have traditionally been preserved as the sites for their highest courts and ceremonies, despite (a westerner would say) their considerable economic potential.

> Kaya forests are attached to and protected by specific communities and clans. The very existence of the kaya lies

> not so much in effort to control the cutting of trees or the sale of the land, as it lies in the continuation of the social existence of the communities and the forest's cultural significance within these communities.[21]

But kaya faced a challenge from competing traditions within the Mijikenda communities. Matrilineal, clan-based land inheritance practices of Muslim Mijikenda threatened to fragment ownership of the kaya and extend ownership to people outside of the immediate community, while the communal property traditions of non-Muslim Mijikenda endeavored to limit access to the kaya based on membership in the community. The solution was the establishment of several kaya as national monuments by the government of Kenya, which legally preserves traditional uses regardless of ownership. For now, two traditional relationships to the land and modern regulatory methods seem to coexist effectively.[22]

Traditional organizations also play an environmental protection role in the Republic of Guinea in West Africa. State administrations and more recently international agencies have supported hunters' brotherhoods in the defense of national parks. As Melissa Leach notes:

> The new policy . . . is to have no park guards, and to work instead through hunters' brotherhoods, now newly respected for their knowledge and authority as custodians of the bush. Working through a 'traditional organization' supports the self-representation of donors and governments as part of a new era of conservation and 'sustainable development' which is 'participatory' and respectful of 'culture' and 'tradition.'[23]

This has introduced some conflict with the increasing hold of Islamic Imans in the area as well as with local agents of the national forest service. Leach also points out that the hunting brotherhoods' new role reinforces a "particular version of masculinity—the macho, gun-phallus-bearing killer" with its need to "control female speech and sexualitiy."[24]

2.4. Aboriginal Values

A final example of traditional values and attitudes with respect to the environment concerns aboriginal populations, found throughout the world, who were once or still are semi-nomadic. Most of these communities engaged in hunting and fishing of species upon which they relied for their survival. Their views of the species, however, were not instrumental. Indeed, the prey hunted "gave up its life" for the hunter, but expected respectful treatment. More an ecosystem than biocentric perspective, aboriginal spirituality saw intrinsic value in both nature and humans, and saw humans as responsible for maintaining a balance of life. Indeed, a recent study argues that traditional American Indian cultural world-views expressed "a higher regard for non-human natural entities and nature as a whole" as compared to the dominant Euro-American perspective.[25]

Lest we over-generalize, however, it is important to note that aboriginal belief systems may also lead to practices that alter the natural landscape for human use. To support their imperial systems, the Inca, Maya and Aztecs mined and refined metals, and built roads and cities. The Inca altered the Andean landscape by terracing, and built irrigation systems to create new agricultural land, while the Aztecs overcame the problem of semi-arid lands by constructing floating platforms in lakes for the planting of food crops.[26]

3. ECONOMIC DEVELOPMENT AND SOCIAL CHANGE

The types of economic activities that take place in nation-states have direct bearing on environmental conditions as well as the policy responses to them. Governments in LDCs often devise explicit development strategies for marshalling the physical, financial and human resources of a nation. Collectivization of agriculture and rapid industrialization—guided by Leninist and Stalinist versions of Marxism in the Soviet Union and Maoism in China—transformed the economies, altered social structures, and scarred the environment of vast regions. The export led growth strategies of Japan after World War II, and the East Asian "tiger" economies in the 1960s and 1970s, although oriented toward industrial production for global markets, also achieved rapid socio-economic change and introduced new sources of industrial pollution.[27]

In such cases states are typically trying to achieve specific economic, political and social goals, including "catching up" with more developed countries that they see as competitors or threats.[28] Post-revolutionary and authoritarian regimes legitimize themselves by protecting the nation from its enemies, and setting and achieving socio-economic objectives. Whether capitalist or socialist, twentieth century development strategies have shown a politically driven bias toward rapid growth, and have rarely considered the environment. Most development strategies have been partially successful, as industrialization has become a global phenomenon. But the gulf between economically developed countries (EDCs) and lesser developed countries (LDCs) has grown since the end of World War II.

EDCs followed one of several capitalist routes to development, which involved elaborate transitions from agricultural to industrial economies and when substantial numbers of people began to produce for markets, they became increasingly competitive on a global basis. Both increased exploitation of natural resources and development of industries caused migration to cities, which broke up tribal or village communities, weakened the traditional family structure, emancipated individuals, and brought about cultural changes.

Walt W. Rostow, whose *The Stages of Economic Growth*[29] summarizes the capitalist development process, described the "preconditions for the takeoff" of modern economic growth, which included incorporating the insights of modern science into more rational agricultural and industrial production, expansion of world markets, and appearance of entrepreneurial elites. Other preconditions included political consolidation and development of transportation and communications infrastructure indispensable for economic growth. At the take-off stage, forces for economic growth overwhelm remaining barriers of traditional society, although this process usually took generations. The final stage was the development of a highly differentiated economy. Of greatest importance, it was simultaneously a stage of "mass consumption," involving the vast majority of the population as participants in the urban-industrial system.[30]

Communist countries followed a different developmental path of two initial stages. First, in the initial years after their assumption of power, the regimes engaged in consolidation of power and economic reconstruction. Internal conditions of post-war communist states were similar: massive destruction as a result of war, extensive poverty and malnutrition, general disorder in industry and agriculture and governmental paralysis. The next principal period was the stage of centralized planning, during which a series of Five-Year Plans were implemented, the first initiated by Stalin in 1928. The Five-Year Plans called for strict centralized organization in all phases of the economy to achieve rapid economic growth (and to surpass that of capitalist rivals).[31] The result was a command economy, with all major economic decisions made by the state and not the market.

Nation-states that became independent after World War II were influenced by both models. Because most had been colonies or dependencies of western imperial powers and remained suspicious of neo-colonialism, the market-based model was less attractive than economic stimulation under the aegis of the state. Also, the raft of nationalist parties holding monopolies of power in newly independent states sought to continue that power, which challenges from market forces would disturb. Most of these newly-independent states sought to direct economic growth, particularly through import-substitution industrialization (ISI) policies.

Whatever the course taken by nations, rapid economic development threatened environments globally. We turn briefly to modernization and political development, and then to dependent development, as different ways in which choices were constructed for LDCs.

3.1 Modernization and Political Development

The "modernization school" of analysis arose in the 1950s and 1960s in order to explain the rise of the industrialized nations of the West and to guide the evolution of post-colonial states in Africa, Asia, and Latin America. As one of the leading theorists of modernization, Daniel Lerner, analyzed the process, modernization involved "psychic mobility" or "empathy,"—"the capacity to see oneself in the other fellow's situation."[32] Modernization theories were dichotomous and posited the transformation of primitive, agricultural, traditional, sacred communities into modern, industrial, secular, universalist state systems. The linear, stochastic, and ethnocentric nature of the model exposed it to early critical commentary.

Students of comparative politics in the 1960s began to think of political development as a process independent of, although obviously affected by, the stages of economic modernization. Sensitive to the ethnocentric charge against modernization models, they attempted to create concepts that could be used cross-nationally, without cultural bias. The most popular approach was structural-functionalism, which was built upon systems theory—the belief that each country could be seen as a system including inputs, outputs, a conversion process (the government), and a feedback mechanism.[33]

This approach explained environmental policies through the inputs or demands by the general public or interest groups for environmental protection. The demand would be processed by government and turned into policies, specific courses of action, such as regulation of vehicle emissions. Behind the political development profile of any nation, however, were certain direct linkages to economic growth.

3.2 Uneven Development and Dependency

Both modernization and development approaches to political and economic change were established in the particular historical circumstances following World War II: the United States was the most economically developed and politically secure superpower, facing a rising and competitive power center in the Soviet Union, and an array of states, including the plurality of less well developed countries in Asia, Africa, and Latin America (then collectively called the "Third World"), presumably in search of the best development strategy to bring them wealth and power.

It was obvious to critics of both modernization and development theories, however, that the process of development for LDCs was dependent on external factors. As Andre Gunder Frank and others pointed out, the structures and dynamics of the world economy, controlled from its centers (especially the United States), locked peripheral societies into an unyielding downward spiral of exploitation and poverty.[34] Imperialism and neo-imperialism had assigned LDCs a subordinate position in the international system—supplying cheap raw materials and labor, and outlets for the profitable investment of surplus capital.

Precursors of dependency theory, especially the analysis done by Argentine economist Raul Prebisch for the United Nations Economic Commission on Latin America, identified industrialization as the key to overcoming dependent underdevelopment, and prescribed ISI as a necessary first step. Brazilian economist Theotonio dos Santos, informed by Marxian analysis of U.S. hegemony in Latin America, argued that dependent states could never develop as part of a capitalist world system.[35] The economic, social and political structures of LDCs were permeated by the interests of the capitalist center. Even ISI left the developing state and its industrializing sector reliant on multi-national corporations (MNCs) for capital, technology, and managerial expertise, and thus enhanced exploitation. Leaders of developing states colluded with MNCs to maintain their privileges and power; both native industrialists and MNCs used the weak state but for different ends, which were equally degrading of popular forces and the environment. These arguments resonated strongly in Latin America where U.S. economic hegemony was manifest and the support by the U.S. government and U.S. based multi-nationals for authoritarian regimes was an open secret.

Beginning in the 1970s, an alternative model of the possibilities for LDC development under capitalism emerged. Initially conceptualized by Brazilian sociologist Fernando Cardoso as "dependent development,"[36] the argument was made that LDCs could achieve sufficient autonomy in the world capitalist system to attain their own development objectives. Central to this possibility was the development of a triadic alliance among the state and indigenous and transnational capital. This approach connected with the view of world-systems theorists such as Immanuel Wallerstein that some peripheral societies could move to intermediate positions in the global economy, where their economies extracted surplus from the periphery while yielding surplus to the core.[37]

What is clear from this review is that although LDCs had several choices when selecting strategies for economic development, they were limited by global circumstances of political and economic power. When making choices, environmental effects were not considered. Instead, decision-making on economic strategies reinforced the dominant environmental perspective of industrialized western nations, or the cold war alternatives of Soviet or Chinese style state socialism. To the extent alternatives to capitalist modes of development existed, they came at a rather high price. Through revolutionary means states might extricate themselves from the capitalist world-system—as Cuba did in 1959—but in the process they isolated themselves from the capital, technologies and markets needed for effective "dependent development." And, as noted above, environmental considerations were no more a part of Cuba's Soviet-supported development strategy than they had been of its colonial or pre-revolutionary economies.

By the late 1980s the competition between liberal modernization theory and Marxian dependency and world-systems approaches seemed to lose its relevance, on both the theoretical and practical levels. The debt crisis of the 1980s and the demise of the Soviet Union in 1991 brought neo-liberalism to the newly democratized states of South and Central America. Almost everywhere, ISI and socialist development strategies were abandoned.

In 1998, anti-imperialism in Latin America was revived with the election of Hugo Chavez as president of Venezuela. Chavez's strategy to unite Latin American economies against U.S. economic domination recalls the economic nationalism of ISI and the radical socialism of some versions of dependency analysis. But the strategy depends on the wealth and political leverage derived from Venezuelan oil, not the decoupling of Latin American resources from the global, industrial capitalist economy.

World-systems theorists have found environmental concerns more amenable to their critique of capitalist development. Writing in 1999, Wallerstein associated the capitalist imperative to expand with environmental degradation, identifying the solution to global environmental crises as the end of the capitalist world-system.

> The environmental dilemmas we face today are directly the result of the fact that we live in a capitalist world-economy. . . essentially because capitalists in this system succeeded in rendering ineffective the ability of all other forces to impose constraints on their activity in the name of values other than that of the endless accumulation of capital.[38]

Nevertheless, Wallerstein does not reject the value of technological modernization, decrying an "unfortunate tendency to make science and technology the enemy, whereas it is in fact capitalism that is the generic root of the problem."[39]

4. A "NEW" ENVIRONMENTAL PARADIGM?

Whether capitalist or Marxist in origin, development strategies of the twentieth century were derived from ideologies of industrialism that revered technology and material abundance over nature. As early as 1933, Aldo Leopold observed:

> As nearly as I can see, all the new isms – Socialism, Communism, Fascism . . . outdo even Capitalism itself in their preoccupation with one thing: the distribution of more

> machine-made commodities to more people. Though they despise each other they are competitive apostles of a single creed: *salvation by machinery.*[40]

The negative marks of the industrial revolution on environments prompted the development of conservation movements in many countries at the end of the nineteenth century. Simultaneously, U.S. Progressive Era reforms produced a wave of regulation, expanded the powers of the federal government over the economy and the states, and established legal limits on corporate behavior. But the contemporary environmental policies tended to be localized and did not affect the organization of power in any state. The environmental movements that developed in post-industrialized societies in the 1960s and 1970s (discussed in the following chapter) were a different story. They were the product of ideas, values, attitudes, and opinions on human-environment relationship, some original, others of long-standing but newly introduced to the mainstreams of academic and political discourse. Many scholars believe that the coalescing constellation of values and attitudes on the environment represents a new environmental paradigm. We explore this contention by treating the creation of environmental philosophy, examining changes in popular values and attitudes, and examining variation in beliefs cross-nationally.

4.1 Creation of Environmental Philosophy

Since the late 1980s, new issues and topics in moral philosophy have caused scholars to re-examine the sufficiency of dominant philosophical perspectives regarding the environment. The new topics include animal rights and liberation, species and ecosystem protection, eco-feminism, and deep ecology among others. Although it is difficult to establish cause and effect, these newly synthesized ways of thinking about the environment are associated with new movements that have affected values and attitudes, and in some cases policies. While there are several definitions of the new environmental philosophy, a relatively comprehensive treatment sees it as composed of three fields: (1) environmental ethics, (2) radical ecology, and (3) anthropocentric reformism.[41]

Environmental ethics challenges norms putting humans at the center of the universe, and argues that the "right thing to do" is to extend moral considerability to non-human entities. Most environmental ethicists include non-human but sentient animals as objects worthy of moral consideration. Yet they disagree whether to treat non-sentient but living entities, such as trees and grass, as of intrinsic value, and this disagreement extends to inorganic matter that is part of the ecosystem. Too, there is disagreement about what justifies treatment of anything as of intrinsic value.

The second field of environmental philosophy—radical ecology—doubtless is most controversial. It includes eco-feminism and deep ecology among other strands. A tight definition of radical ecology holds that humans are simply interdependent parts of a complex biosystem in which they are equal (and not superior) to each and every other species. The environmental crisis is explained by anthropocentrism, the human tendency to treat all else as instruments to satisfy human pleasures.[42]

The third field of environmental philosophy—reformism—is most relevant to policy and also least controversial. It suggests that serious problems of the environment such as air/water pollution, deforestation, biodiversity loss, and climate change are the result of "ignorance, greed, illegal behavior, and shortsightedness."[43] They can be remedied through changes in the law, by increased education, and through wise use of resources. No change in the human tendency to adopt instrumental views toward non-human entities is needed, but human behavior must be altered.[44]

Politically, what these three tendencies have in common is the linkages they make between environmentalism and critical perspectives on other issues, including civil rights, human rights, indigenous rights, economic and social inequality, and gender discrimination.

For example, the link made by eco-feminists between feminist thought and environmentalism is not new, but does constitute freshly articulated theoretical perspective on the environment. Eco-feminists such as environmental historian Carolyn Merchant argue that the analysis of environmental problems based on changing modes of production, ideas and perceptions is enhanced by the addition of gender analysis. Women's roles in production, as societies developed from hunting and gathering, to agrarian, to industrial, were distinct from men's and therefore produced different attitudes toward the environment. Additionally, the role of women in reproduction, at each stage of development, is unique and critical to human effects on ecology. Reproduction is understood in two ways: the biological processes of birth and nurturing; and the educational processes of childrearing, teaching life-skills and socialization.[45] An "overtly feminist perspective on the environment," therefore, is not new. Although women linking environmental, human rights, and a distinctly feminine perspective on nature and society is most recently associated with new social movements such as the anti-nuclear movement in Great Britain, and grassroots activism among indigenous women in Central America and southern Mexico, feminist advocacy of environmental issues is of long-standing and, it can be argued, continuous. Lydia Adams-Williams, an early twentieth century US conservation writer suggested:

> As it was the intuitive foresight of [Isabella of Spain] which brought the light of civilization to a great continent, so in great measure, will it fall to woman in her power to educate public sentiment to save from rapacious waste and complete

exhaustion the resources upon which depend the welfare of the home, the children and the children's children.[46]

More recent and dramatic examples exist of new perspectives linking the environment to other social issues. In these cases scholars deduce philosophical change from examples of new practices that link the environment with human rights.

Since the 1950s deforestation by the timber industry and the clearing of land for cattle ranching have denuded much of the Mexican state of Guerrero of its forest cover, degrading the soils of peasant *ejidos* (traditional communal agricultural lands until recently guaranteed by the Mexican government) and jeopardizing water supplies. The federal government, although officially committed to providing financial, technical and infrastructural support to the *ejidos* has largely favored the interests of loggers in the building of roads and the enforcement (or lack thereof) of environmental regulations and development policy. State government in federalist Mexico has been complicit with local and national political bosses in serving the interests of industry.

Early peasant protests in Guerrero were essentially about the challenges of making a living off the land and a political system that was either unresponsive or violently hostile to the needs of small farmers. By the 1990s peasant protests were taking on an expressly environmentalist cast. The Organization of Peasant Ecologists of the Sierra de Petatlán and Coyuca de Catalán, formed to call the government's attention to the environmental effects of uncontrolled and often illegal logging on the hillsides of the region, and to demand better and better-enforced environmental policies. In 1999, one member was killed in an attack by the Mexican Army, and two of the organization's leaders were taken into custody where they were reportedly tortured and eventually convicted of weapons charges on the basis of coerced testimony. Pressure on the Mexican government by international environmental and human rights organizations led to their release from prison after one year for "humanitarian reasons."[47]

The linkages of environmental to other social and political issues are particularly apparent in the field of environmental justice. "While the nerve-centers of Deep Ecology are in the wild, environmental justice is firmly rooted in human habitations. The threats it fears are toxic waste dumps and landfills, the excretions of affluence that have to be disposed of somehow, and somewhere."[48]

In the U.S. the environmental justice movement arose from high profile cases of victimization. Perhaps the best known is the 1978 Love Canal incident in which residents of a lower-middle class housing development in New York State were able to trace elevated occurrences of certain diseases and birth defects to toxic wastes dumped in the building site over decades by the Hooker Chemical Company.[49] In the 1980s, empirical research by non-governmental organizations established correlations among low income, racial and ethnic segregation, and increased exposure to environmental hazards. Since the 1980s environmental justice and environmental racism have become small but important parts of academic and policy debates on racism and inequality, as well as on urban air and water quality and waste disposal.[50] Many activists see the correlations among poverty, ethnic minority status, and increased exposure to environmental risks played out globally as well, charging that EDCs export their toxic wastes and polluting technology to LDCs. Such linkages make up an important part of the appeals of international environmentalist and anti-globalization movements.[51]

4.2 Changes in Values and Attitudes

One suspects that a small minority of citizens, and then only in post-industrialized societies, has embraced all three strands of the environmental philosophy. Yet we have evidence that values and attitudes toward the environment have been changing in fundamental ways, represented by the "post-materialist" concept, which is part of a far broader value shift represented by postmodernization.

The chief architect of this analysis is Ronald Inglehart who, beginning in the late 1980s, updated and revised modernization theory in four ways. First, he stated that change is not linear but instead reaches points of diminishing returns and moves in new directions. Second, instead of economic or cultural determinism, Inglehart posits that relationships among the economy, culture, and polity are mutually supportive—a pattern of reciprocal causal linkages. Third, he rejects the ethnocentric perspective of those who equate modernization with westernization, instead viewing it as a global process. Finally, he disputes the assertion that democracy is inherent in modernization, arguing that it becomes more likely as societies become post-modern.

Inglehart acknowledges the extent to which the concept of postmodernism may be a rejection of modernity and a revalorization of tradition. His work emphasizes the "rise of new values and lifestyles, with greater tolerance for ethnic, cultural, and sexual diversity and individual choice concerning the kind of life one wants to lead."[52]

Why the post-modern shift occurs requires investigation of broad-ranging transitions from agricultural to industrial and on to post-industrial societies. The trend toward bureaucratization, centralization of power, and government ownership and control has reached its functional limit and begun reversal, in Inglehart's view. Norms and expectations underlying human behavior have changed: from the politics of class conflict we move to "political conflict based on such issues as environmental protection and the status of women and sexual minorities." The origin of this value shift is the welfare state and the end of scarcity. Now, increasingly (at least in the West), people grow up with the feeling that one's survival can be taken for granted. The change is gradual and not abrupt for two reasons. First, scarcity—individual priorities reflect the socioeconomic environment. Second, socialization—a time lag is involved because basic values reflect the conditions prevailing during one's pre-adult years.

4.3 Evidence of Post-Materialist Values and Attitudes

A number of recent attitudinal studies have documented the existence of post-materialist values and attitudes. For example, Lester Milbrath contrasts the new environmental paradigm (NEP) with the dominant social paradigm in two respects: advocacy (as compared to resistance) of social change, and a high value given to a beautiful environment (compared to the value of material wealth). Analyzing national attitude survey results, he identifies five attitude groupings in the American context:[53]

The vanguard, supporters of the new environmental paradigm	15-17%
Nature conservationists who call for a balance in values	7-8%
The rearguard, supporters of the dominant social paradigm	15-17%
Environmental sympathizers with both sets of beliefs	60%
Deep ecologists, neither involved in politics or political reform	1-3%

From this analysis, it seems clear that less than a majority support the new environmental paradigm; moreover, more members of the mass public appear to have adopted it than elites. ("Sympathizers are divided, but NEP sympathizers are more numerous than members of the vanguard.") Nevertheless, it represents a new axis in American public life, on occasion more important than the old left-right dichotomy.[54] The NEP emphasizes values and issues of increasing importance, such as valuation of nature, environmental protection, generalized compassion, handling risk, limits to growth, need for a new society, and need for a new politics.

Some information is available on value shifts in other countries. Pierce, Lovrich, Tsurutani, and Abe conducted a systematic survey (via mail) of samples of activists, elites, and citizens of two counties in the United States and Japan. They found a causal connection between environmentalism and post-materialist values, related to the security and affluence of post-industrial societies. In general, in both Japan and the U.S., there was support for NEP at all political levels, with activists and elites more supportive than the public. In the U.S., there is greater activist and elite constraint (integration of beliefs) than among the public; in Japan, the reverse obtained. The chief finding was that there were fewer Japanese with post-materialist values than in other western nations, attributable to three factors: (1) the Japanese environmental movement is different from that in the U.S. The former is victim-oriented and developed in response to coastal pollution; the latter aspires to preserve the environment in its natural state. (2) There are as many supporters of NEP in Japan as in the West, but supporters are less likely to share attributes of typical post-industrial advocates. (3) The substance of NEP support in Japan represents a less radical departure from the dominant social paradigm than in the West. The Japanese see themselves as far more highly integrated with nature. The implication is that support for NEP is culturally based and will differ from country to country.

A somewhat more impressionist study, using the Q-sort methodology, suggests that Indian elites share attitudes respecting protection of biodiversity, respecting all living things, and the need for energy efficiency. One section, which author Peritone calls the "greens," opposes nuclear power development, biotechnology, and international development planning, but supports grassroots economic development that is sustainable and democratic.[55] This group stands in contrast to "ecodevelopers" who are political realists strongly promoting economic development. Yet they also oppose biotechnology and nuclear power. Finally, the "managers" give priority to human needs and rational management of environmental processes. The "greens" represent a larger number from this small sample of 34 leaders than the other two leadership approaches.

Scholars have also presented broader interpretation of the change in traditional and religious values, in support of environmental protection objectives, in different regions. For example, Hsiao et al. present "frames" of how people in Taiwan, Hong Kong, and the Philippines observe the world, form and develop identities, and determine what is sacred, what is profane. As environmental movements have established roots outside the West, they have incorporated traditional religious and cultural beliefs. Religious sanctions have been invoked to legitimize protest activities. Confucian values (in Taiwan and Hong Kong) have been reinterpreted for environmental purposes, extending to the use of Confucian familism, the role of funerals, and the status of motherhood. Finally, movements have invoked indigenous traditions to protect communities and native lifestyles.[56]

4.4 Variation Cross-Nationally

Inglehart's study is the first to systematically examine the degree of support for post-materialist values cross-nationally. From 1990 to 1993, he directed surveys in 43 societies,

> representing almost 70 percent of the world's population and covering the full range of variation, from societies with per capita incomes as low as $300 per year to societies with per

> capita incomes as high as $30,000 per year, and from long-established democracies with market economies to ex-socialist states and authoritarian states.[57]

The evidence from these surveys supports two quite different findings. First, societies with high levels of economic development also have high levels of postmaterialist values. Second, societies with high rates of economic growth show large differences between the values of younger and older generations,[58] because the pre-adult experiences of the younger generation have been more secure than those of the older generation. Although most of the countries with high levels of postmaterialist values are western nations, economically developed Asian countries, such as Japan and South Korea, show similar levels.

Inglehart argues that the rise in postmaterialist values helps account for the important rise in salience of environmental issues. In advanced industrial nations, he maintains that environmental protection is mainly a postmaterialist concern. In developing nations, such as China and Mexico, however, "air pollution and water pollution levels are far worse than in advanced industrial societies, posing immediate problems to health. In such settings, environmental protection is not a quality of life issue, but a matter of survival."[59] For these countries, environmental protection is as likely to be supported by materialists as postmaterialists. One example of public response to environmental issues is seen in recent studies of China. There, survey research on the environment is relatively recent, dating only from 1990. One sees a variety of survey research products. Some are commissioned by government agencies, such as the State Environmental Protection Administration (SEPA), some by universities or NGOs; many use samples of convenience. Ultimately, the data are non-comparable. Finally, we lack longitudinal data that would indicate *change* in public opinion over time. Nevertheless, a dozen reports provide sufficient information to make several preliminary observations.[60]

First, there is a growing awareness of environmental problems. Among the list of problems for which respondents have been asked to make assessments, items affecting species and ecosystem degradation (for example, desertification, chemical pollution, reduction in biodiversity) are regarded as equal or greater in seriousness to pollution of air, water, and land.[61] The various reports, however, indicate that the consciousness of environmental degradation may be superficial, given lack of uniform treatment of environmental issues in schools. Awareness also varies by region (rural versus urban) and expresses the not-in-my-backyard (NIMBY) phenomenon: people are more likely to be aware if they have direct experience of an environmental problem.

Second, most respondents assign a lower priority to environmental problems than to other issues such as unemployment, overcrowding, and educational quality, with the exception of young people who give environmental protection a greater value than the middle-aged or old.[62] Moreover, survey research indicates an unwillingness of respondents to make the trade-offs necessary to improve environmental conditions by, for example, slowing economic growth. Reasons for this lack of commitment have not been specified through intensive field research, but one plausible explanation is "lack of trust in government officials' commitment on matters pertaining to the environment."[63]

The research to date fails to document a strong influence of public opinion on governmental performance. One commentator notes: "Instead of policies being informed or influenced by public opinion, it is the public's own environmental perceptions of the environment that are being shaped by state policies propagated by the media."[64]

Finally, survey research in China indicates that most respondents continue to hold anthropocentric views concerning nature. A recent survey in Guangzhou found that a quarter or fewer respondents agreed to tenets[65] of the New Environmental Paradigm (NEP), a lower percentage than found in Europe, North America, and Japan.[66] Although this study did not identify the characteristics of those adopting NEP, we suspect they are young, well-educated, and hold professional occupations.

These studies in China present an interesting contrast to Inglehart's broader comparative work. He finds the highest levels of support for environmental protection in the Nordic countries and the Netherlands; these nations also have the most postmaterialist publics in the world.[67]

Altogether, these different studies make the argument that as nations develop economically, values and attitudes of citizens become less materialist and more likely to accommodate quality-of-life concerns, among which is environmental protection. An important caveat, however, is that citizens in developing nations who experience environmental pollution are likely to have environmental protection attitudes as well, notwithstanding the generally materialist cast of their values.

4.5 Attitudes, Institutions and Development

Studies of postmaterialist values show that environmental issues are more likely to affect political processes in economically developed countries but they do not explain why that happens to everyone's satisfaction. World-system theorists see the apparently positive effects of postmaterialism as little more than a new method for shunting the costs of capitalist production onto less developed countries. Wallerstein argues that greater demands for environmental responsibility in developed countries will amount to very little. Left to their own devices, large capitalist enterprises will not internalize the environmental costs of their production activities. Therefore, citizens in developed countries must pressure states to find ways to compensate. But to bear the costs, states must raise taxes. Since taxes on enterprises will cut into profits and taxes on consumers will be politically unpopular, a third and more likely option is to "do virtually nothing." Alternatively, in a global capitalist system developed countries have another, albeit temporary, solution—to transfer the ecological costs of expansion to the global South. This can be done by exporting wastes and/or demanding slower rates of development and the use of cleaner, more expensive technologies.[68]

Presumably, political modernization (understood as democratization of LDCs) would make it harder for the EDCs to dump their environmental problems in LDCs. But, Wallerstein contends, democratization in LDCs increases demands for material goods making the expansiveness of capitalist production more popular as more people come to demand their rights to higher levels of consumption. As Inglehart points out, however, when people become economically secure and politically engaged, they want not only more production, but also the qualities of life that a clean natural environment provides. "That is, many people want to enjoy both more trees and more material goods for themselves, and a lot of them simply segregate the two demands in their minds."[69]

Two other lines of reasoning disagree sharply with the world-systems theorists, and come to similar conclusions as Inglehart about environmental politics—but for different reasons. These are the Environmental Kuznets Curve (EKC) and Ecological Modernization.

Proposed in 1955 by economist Simon Kuznets, the Kuznets Curve sought to explain the relationship between economic growth and inequality by showing that as per capita income in a society increases, economic inequality at first increases and then declines. The pattern is an inverted "U" with a pronounced upswing in inequality during the early stages of economic growth, followed by a gradual decrease toward a more equitable society. Advocates of free trade and economic growth as paths to environmental improvement have examined cross-national data on specific pollution problems associated with development—for example, sulfur dioxide concentration, urban air quality, and heavy metal contamination of rivers—to reveal an Environmental Kuznets Curve. In brief, the argument states that environmental damage will increase sharply during early stages of economic development (understood principally as industrialization) but once per capita income reaches a certain level (at around the dividing point between lower- and upper-middle income levels shown in Table 1.1) environmental harms will be gradually reduced and environmental improvements will ensue. Free trade and the desire for economic growth are seen as positive forces, as they provide incentives for states to specialize in the types of production they can do most efficiently. Greater economic efficiency will mean less pollution per unit of production. Economic growth will also attract new investment, including transfers of cleaner technologies; and consumers will be able to afford, and will therefore begin to demand, more environmentally friendly products. In short, over time economic growth can be expected to have benign environmental effects.[70]

"Ecological Modernization Theory" also posits positive environmental effects from economic development, but Arthur Mol and others argue that environmental improvements do not have to wait for economic "take off." They theorize a process of "ecological modernization" through which "the growing autonomy of an ecological perspective and rationality . . . has started to challenge the dominant economic rationality."[71] They argue that in the 1970s and 1980s, especially in developed countries, environmental concerns took on an importance in politics separate from and often in competition with economic interests. One important result was the creation of environmental institutions (both governmental and non-governmental), followed closely by the emergence of viable green political parties. Thus, "a distinct green ideology" entered the politics of many developed countries, different from the competing socialist, liberal and conservative ideologies of traditional political parties and policy-making perspectives. Mol posits a diffusion effect from economically developed to developing countries. International agreements, international and inter-governmental organizations, and even multi-national corporations have become increasingly green, and in alliance with more ecologically aware elements of the civil societies of LDCs they spread the process of "eco-modernization." Evidence is offered from global measures of the growing number of environmental laws, agreements and organizations, and the technological innovations spread by foreign direct investment. Perhaps the clearest example of vertical diffusion is from the European Union to its 10 new member states, which entered the confederation in 2004. Old members, particularly Germany (making up one-third of the EU's economy) are post-industrial and display postmaterialist values. The EU now is spreading these values to central and eastern European member states. Although the effectiveness of ecological modernization in improving environmental conditions in EDCs or LDCs is still unproven, Mol argues that the institutional changes along with the changes in attitudes and awareness are propitious.[72]

4.6 A Note on Sustainable Development

Since the late twentieth century, sustainable development has become the key concept of environmental politics. In the 1987 Brundtland Report of the UN's World Commission on Environment and Development, sustainability was defined as meeting "the needs of the present without compromising the ability of future generations to meet their own needs."[73]

The concept is ambiguous and has more than two dozen contrasting definitions. A weak definition may mean little more than paying heed to future generations when planning change. A strong definition, such as that by Mazmanian and Kraft, implies "important change of values, public policy, and public or private activity that moves communities and individuals toward realization of the key tenets of ecological integrity, social harmony, and political participation."[74] In this strong sense, sustainable development is among the values included within Inglehart's postmaterialism.

Sustainable development was given operational force in Agenda 21 of the 1992 Rio de Janeiro UN Conference of Environment and Development. By this we mean that the concept was translated into a set of policy strategies, and linked to other objectives such as poverty alleviation and participatory practices in environmental management.[75] But sustainability remains a notoriously fluid concept. As a guide to institutional design and political practice it allows the integrated pursuit of economic growth and environmental preservation.[76] But it is up to states to design the institutions, promulgate the policies, and calculate the trade-offs between growth and the environment, and between present and future needs. Lafferty and Meadowcroft conclude that "the expanded normative conceptual scope of sustainable development has not only been taken seriously within high-consumption societies, but that is has also given rise to new constellations of political forces within individual countries . . ."[77] Najam observes a similar impact on the attitudes of LDC policy-makers.[78] However, research to date has not established a relationship between countries' adoption of the concept and improvement in environmental outcomes.

5. SOCIAL ORGANIZATION AND STATE DECISION-MAKING

In this final section we move beyond value and attitude states to consider the relationships between state and social groups, a discussion which continues in the following chapter. Individuals tend to associate with like-minded persons, and form groups to advance their values as well as their interests in the larger society. The relationships between groups and states vary cross-nationally. Here we discuss relationships between economic groups, or market society, and the state; in the next chapter we examine environmental groups and movements.

In broad terms, groups may organize freely and stand in apposition to the state—a system called *pluralism*—or the state may organize groups and direct their energies, in a system called *corporatism*. Our discussion focuses on organized labor and business, and whether they compete freely or are directed by the state largely is explained by the period when industrialization occurred in the nation and the strength of the state at that time. Most pluralist systems saw the state developing strength after industrialization had occurred, and both labor and business interests had organized collectively. This was the pattern in both Great Britain and the United States. Most corporatist systems saw the development of a strong state first, whereupon industrialization occurred and business and labor interests organized, but under the influence of the state. This was the strategy used by the German Second Reich and the fascist regimes of the 1920s-40s in Italy and Germany, and somewhat differently the military and populist regimes in late developers such as Mexico under Lázaro Cárdenas, Argentina under Juan Perón and Brazil under the military regimes of the 1960s and 1970s.

Corporatism thus has meaning in contradistinction to *pluralism*, which is a bedrock idea in the formation of civil society. The pluralist approach conceives of the market (and the broader society) as configured into different interest groups and associations. The groups are voluntary associations, free to organize and gain influence over state policy commensurate with their political resources. The state may or may not intervene regularly in the market, but it serves as a real or potential arbiter of market conflict. In pluralist systems, the state appears to be weak, in the sense that it can be penetrated by the strongest interest groups, frequently by dominant business groups and coalitions. Thus, the direction in which the state is influenced is the outcome of interest group conflict.

Corporatist systems,[79] on the other hand, are those in which the state may be strong enough to formulate economic policy without becoming captive to rent-seeking groups. Whether strong or weak, the state is actively involved in the market, and attempts to influence the use of both public and private resources in accord with a vision of how the industrial structure of the country should be evolving. Unlike pluralism, where there may be open competition between groups in society and where groups are potentially equal and have access to centers of political authority, corporatism observes an hierarchy of interest representation and unequal access that is institutionalized. Both business and labor unions are hierarchically ordered, which has obvious implications for environmental movements challenging business power. The corporatist perspective does not ignore the development of new social forces but denies them autonomy.

Corporatist institutions configure a system of interest representation in which a small number of strategic actors (invariably capital, sometimes labor), which are organized in peak associations, represent most of the population in an "encompassing" fashion. In some cases political parties may act as intermediary institutions in connecting peak organizations to the state; what might be called "party corporatism." This was the case in Argentina under Juan Perón whose *Justicialista* Party controlled the country's largest labor confederation, using it as a bulwark against the resistance of agricultural producers to import substituting industrialization. This was also true of Mexico under the *Partido Revolucionario Institucional* (PRI, 1929-2000), where the country's major labor and peasant confederations made up two of the three organized branches of the party.[80] Current forms of corporatism, commonly found to varying degrees in Western Europe, tend to be of a milder, more democratic form. In France, Spain and Norway for example, labor and business leaders are consulted by state and government officials on general economic policy. Nevertheless, business and labor are provided access to policy makers rather than having to vie for it through lobbying and campaign contributions. Pluralist institutions, on the other hand, shape a large number of atomistic interest groups engaged in a competitive struggle to influence national policy. Two recent studies suggest that corporatist institutions are more likely to reduce pollution levels than pluralist ones.

Crepaz asks whether corporatism retards pollution, because the inclusive structure of corporatism allows the internalization of externalities. This hypothesis is based on Mancur Olson's argument[81] that the more encompassing organizations become, the more their interest and the general interest converge. Corporatism, Crepaz suggests, uniquely resolves collective action problems, limits transaction costs, and reduces uncertainty, and this enables it to tackle environmental problems.[82]

The test Crepaz devises measures a cross-national panel of 16 members of the Organization of Economic Cooperation and Development (OECD, all of which are "high income" countries) at two points in time—1980 and 1991. He finds that those nations with a high degree of corporatism had reduced levels of sulfur oxide, nitrogen oxide, particulates, and carbon dioxide, four of the five pollutant sources. He explains this by institutional structure: that a clean environment is a public good. When sources of pollution are centralized (from nationalized, industrial plants), then corporatist institutions are more effective in their reduction than pluralism. Moreover, as a number of studies have established, corporatist societies are more likely to see negotiated voluntary agreements on pollution control struck between business and the state.[83]

Lyle Scruggs, too, argues that there is a robust positive relationship between corporatist institutions and national environmental performance. Just as corporatist institutions promote economic public goods (e.g., wage restraint), so they provide non-economic public goods by overcoming collective action problems characterizing environmental sustainability.[84]

Scruggs' study compares 17 OECD nations (in North America and Europe) from the 1970s to 1990s, defining his dependent variable, environmental performance, as the result of human responses to human-induced environmental pollution problems such as emissions of sulfur dioxide, nitrogen oxide, and municipal waste. His primary explanatory variable is a high level of corporatism, and he finds that it has strong, stable, and statistically significant effects on environmental performance.

Why might corporatist institutions be more conducive to environmental regulation of production than pluralist ones? Scruggs suggests several reasons. First, under corporatism, the government retains the threat to use direct regulation. Second, monitoring and enforcement, necessary to effective environmental regulation, are more acceptable when there is a history of producer-government trust. Third, corporatist institutions seem to have a better ability to pursue public goods than do pluralist ones, because of three factors: (1) national peak associations have power over local units and can reduce parochial interests and avoid policy paralysis, (2) corporatist arrangements have better schemes to compensate losers with economic adjustments, and thus socialize the distributional costs of environmental policies; and (3) producers in corporatist states are active agents in resolving environmental problems.

Another recent examination by Dryzek et al., *Green States and Social Movements*, takes this analysis one step further. Comparing four EDCs—the United States, Great Britain, Germany, and Norway—the authors find a strong relationship between the degree of inclusion of social movements (in this case environmentalism) and changing policy directions of the state. Norway, for example, is noteworthy because it is the most inclusive of the four states, and the social movement has influenced state policy through linking its goals to state objectives, particularly ecological modernization.

Most scholars studying the impact of different models of state-business relations have worked in the capitalist environments of North America and Europe, and one should not assume that the concepts are of immediate applicability to developing countries such as China and India. Yet a number of scholars have tussled with their applicability, given the increase in number of social groups following industrialization and some political liberalization.[85]

In China, there has been a veritable explosion of economic organizations in the 1980s and 1990s, matched by a proliferation of environmental groups in the 1990s (treated in chapter 3). During the 1980s, as the economy liberalized, the government created a large number of business associations. The structure of the non-governmental associations (*minjian xiehui*) resembled the corporatist type found in Taiwan at that time (and previously in Japan), for they were officially registered and only one organization represented each sector. The party-state still controlled structure and personnel of most associations in the 1990s, but some had demonstrated independence. This prompted a United Front Department official to opine that "the non-public economic sector . . . has started to seek the political means to protect its own interests."[86] Studies of labor organizations also pointed to evidence of autonomy in unions.[87]

The degree in autonomy of economic organizations vis-à-vis the state is the crucial issue, and most analysts have cautioned that while the communist party remains in charge of the state, it will patrol limits to the independence of social groups. In recognition of this, the hybrid concept "state corporatism" is used by several scholars to capture the growth of social groups consequent to economic change, within the framework of a Leninist party-state.[88] The applicability even in this context is challenged, because the concept seems unable to explain the state's continued domination of labor, through the All-China Federation of Trade Unions,[89] and state-business relationships in rural areas.[90]

Saich summarizes the general applicability of the corporatism concept to China:

> [C]orporatism as a theory captures well the top-down nature of control in the system and how citizens are integrated into vertical structures where elites will represent their perceived interests. However, such explanations risk obscuring both important elements of change and oversimplifying the complexities of the dynamics of the interaction . . . . New social organizations, for example, can have considerable impact on the policy-making process by retaining strong linkages to the party and state, far more than if they were to try to create an organization with complex operational autonomy . . . . These social organizations with close government links often play a more direct role in policy formulation than in other developing countries as they do not have to compete in social space with other NGOs for dominance and access to the government's ear on relevant policy issues.[91]

During this transitional phase of state-society relations in China, perhaps the most that can be said is tautological: groups with linkages to the state have new avenues for influence and may embed some of their environmental goals in policy.

6. SUMMING UP

Three basic points emerge from the preceding discussion. First is the complexity of the relationship between state and society and the broad variability among nation-states in the ways that traditional beliefs and changing conditions interact to affect environmental politics. Second is that social and cultural change play a critical role in bringing environmental concerns to the national political arena but that the effects of these changes are shaped by the organizational structures, attitudes and political capabilities of the different actors. Third is that although at least some variability in states' responses to environmental issues can be explained by each of the social, cultural, institutional and economic variables mentioned in this chapter we are ultimately drawn to concentrate on political institutions and processes as they respond to and influence changing attitudes and interests related to the environment.

---

[1] A great deal has been written on an emerging "transnational" or "global civil society" in which human rights, democratization and environmental networks shape national policies in ways that are frequently interconnected. See for example, Ann M. Florini, ed.,, *The Third Force: The Rise of Transnational Civil Society*, Tokyo: Japan Center for International Exchange and Washington: Carnegie Endowment for International Peace, 2000.

[2] John Barry, "Sustainability, Political Judgment and Citizenship: Connecting Green Politics and Democracy," in *Democracy and Green Political Thought: Sustainability, Rights and Citizenship*, edited by Brian Dougherty and Marius de Geus. London: Routledge, 1996, 115-31; Peter Christoff, "Ecological Citizens and Ecologically Guided Democracy," in Doherty and de Geus, 151-169; Daniel J.Fiorino, "Environmental policy and the participation gap," in *Democracy and the Environment: Problems and Prospects*, edited by William M. Lafferty and James Meadowcroft, Cheltenham, UK: Edward Elgar, 1996; Bronwyn M. Hayward, "The Greening of Participatory Democracy: A Reconsideration of Theory," in *Ecology and Democracy*, edited by Freya Matthews. London: Frank Cass, 1996, 215-36; Martin Janicke, "Democracy as a Condition for Environmental Policy Success: The Importance of Non-institutional Factors," in Lafferty and Meadowcroft (do you have page numbers for this?); James Meadowcroft, "Deliberative Democracy," in *Environmental Governance Reconsidered: Challenges, Choices, and Opportunities*, edited by Robert F. Durant, Daniel J. Fiorino, and Rosemary O'Leary. Cambridge: MIT Press, 2004, 183-217; Mike Mills, "Green Democracy: the Search for an Ethical Solution," in Dougherty and de Geus, 97-114.

[3] In discussing path dependency, the economic historian Douglass North reminds us that "history matters." While this may seem obvious to most, it is something of which political scientists and policy makers must frequently be reminded. See, D. North, *Institutions, Institutional Change and Economic Performance*. New York: Cambridge University Press, 1990, 100.

[4] For an authoritative discussion of the development of comparative political studies see, James A. Bill and Robert L. Hardgrave, *Comparative Politics: the Quest for Theory*. Columbus, OH: Merrill, 1973.

[5] Gabriel A. Almond and Sidney Verba, *The Civic Cultu re.*Boston, MA: Little, Brown & Co., 1963, 14.

[6] Lester Milbrath, "Culture and the Environment in the United States," in *Environmental Management*, Vol. 9, No. 2, 165.

[7] Quoted in Daniel B. Weber, *John Muir: the Function of Wilderness in an Industrial Society,* unpublished Ph.D. thesis, University of Minnesota, 159-60.

[8] For a concise discussion of "conservationism" and "preservationism" in western environmental thought and policy see, Ramachandra Guha, *Environmentalism: A Global History*. New York: Longman, 2000, chapters 3 and 4.

[9] Significantly, the first product of value extracted from Portugal's American colonies was "brazilwood" (giving the colony its modern name) which produced a coveted red dye. Land thus cleared was later devoted to the large-scale cultivation of sugar cane.

[10] Miguel León Portilla, "Mesoamerica Before 1519," in *The Cambridge History of Latin America: Volume I, Colonial America*, edited by Leslie Bethel. Cambridge and New York: Cambridge University Press, 1984, 3-36; John Murra, "Andean Societies Before 1532, in Bethel, pp. 59-90; H.B. Johnson, "The Portuguese Settlement of Brazil, in Bethel, pp. 249-286; J.H. Elliott, "The Spanish Conquest and Settlement of America, in Bethel, 149-206; John A. Crow, *The Epic of Latin America, 4th edition*. Berkeley, Los Angeles and London: University of California Press, 1992, chapters 1-4; John Charles Chasteen, *Born in Blood and Fire: A Concise History of Latin America*. New York and London: W.W. Norton & Co., 2001, chapters 2 and 3.

[11] "Forest in Cuba: Crazed Heirs of Wooden Treasure," *Granma International* 18 (October 1995).
[12] Manuel Moreno Fraginals, "The Death of Cuban Forests," in *Tropical Rainforests: Latin American Nature and Society in Transition*, edited by Susan E. Place. Wilmington, DE: Scholarly Resources, 2001, 49-55.
[13] Lao Tzu, *Tao Te Ching*, translation by Gia-fu Feng and Jane English. New York: Vintage Books, 1972, 29.
[14] Judith Shapiro, *Mao's War Against Nature*. New York: Cambridge University Press, 2001, 213.
[15] William Theodore de Bary, ed., *Sources of Japanese Tradition*, Volume I. New York: Columbia University Press, 1968, 367; see also Yasuhiro Murota, "Culture and the Environment in Japan," in *Environmental Management*, Vol. 9, No. 2 (1985), 109.
[16] Sarvepalli Radhakrishnan and Charles Moore, eds., *A Source Book in Indian Philosophy*. Princeton, NJ: Princeton University Press, 1960, 91.
[17] Quoted in Ramachandra Guha and Juan Martinez-Alier, *Varieties of Environmentalism*. London: Earthscan, 1997, 158.
[18] See, among others, M. Izzi Deen, "Islamic Environmental Ethic, Law and Society," in *Ethics of Environment and Development: Global Challenge and International Response*, J. Ronald Engel and Joan Engel, eds. Tucson, 1990; A. R. Agwan, *The Environmental Concern of Islam*. New Delhi, 1992; and A. Ba Kader, *Environmental Protection in Islam*. Washington, D.C., 1995.
[19] Zuhair S. Amr and Mahdi Quatrameez, "Wildlife Conservation in Jordan: A Cultural and Islamic Perspective," in Vivek Menon and Masayuki Sakamoto, eds., *Heaven and Earth And I*. New Delhi: Penguin Enterprise, 2002, 175-76.
[20] Yash Tandon. "Grassroots Resistance to Dominant Land-Use Patterns in Southern Africa." in *Ecological Resistance Movements: the Global Emergence of Radical and Popular Environmentalism*, edited by Bron R. Taylor. State University of New York Press, 1995, 161-176; and Ben Wisener. "*Luta*, Livelihood, and Lifeworld in Contemporary Africa," in Taylor, 177-200.
[21] Bettina Ng'Weno, "Reidentifying Ground Rules: Community Inheritance Disputes among the Digo of Kenya," in *Communities and the Environment: Ethnicity, Gender, and the State in Community-Based Conservation*, edited by Arun Agrawal and Clark C. Gibson, New Brunswick, NJ: Rutgers University Press, 2001, 112.
[22] Ng'Weno, 120-34.
[23] Melissa Leach, "New Shapes to Shift: War, Parks and the Hunting Person in Modern West Africa." *Journal of the Royal Anthropological Institute, Vol. 6 (2000): 580.*
[24] Ibid., 592.
[25] J. Baird Callicott and Michael P. Nelson, *American Indian Environmental Ethics: An Ojibwa Case Study*. Upper Saddle River, NJ: Pearson/Prentice Hall, 2004, 1.
[26] Crow, chapters 1-3.
[27] Vinod Thomas and Tamara Belt, "Growth and the Environment: Allies or Foes?" *Finance and Development,* Vol. 34, No. 2, (June 1997), 22.
[28] For otherwise excellent studies of development strategies that provide no discussion of environmental effects, see John Sheahan, *Patterns of Development in Latin America: Poverty, Repression, and Economic Strategy*. Princeton, NJ: Princeton University Press, 1987, and Stephan Haggard, *Pathways from the Periphery: The Politics of Growth in Newly Industrialized Countries*. Cornell University Press, 1990.
[29] W. W. Rostow, *The Stages of Economic Growth*. London: Cambridge University Press, 1963, 5-9.
[30] See Peter H. Merkl, *Modern Comparative Politics*, 2nd edition. Hinsdale, IL: Dryden Press, 1977, 7.
[31] Gary Bertsch and Thomas W. Ganschow, *Comparative Communism*. San Francisco: W.H. Freeman, 1976, 326.

[32] Daniel Lerner, *The Passing of Traditional Society: Modernizing the Middle East*. New York: Free Press, 1958, 50-51.
[33] See, among others, David Easton, *A Systems Analysis of Political Life*. New York: John Wiley & Sons, 1965; Karl W. Deutsch, *The Nerves of Government*. New York: The Free Press, 1963; Gabriel Almond and James Coleman, eds., *The Politics of the Developing Areas*. Princeton: Princeton University Press, 1960; and Gabriel Almond and Sidney Verba, *The Civic Culture*. Princeton: Princeton University Press, 1963.
[34] A. G. Frank, *Latin America: Underdevelopment or Revolution: Essays in the Development of Underdevelopment and the Immediate Enemy*. New York: Monthly Review Press, 1969.
[35] Theotonio dos Santos, "The Structure of Dependence," *The American Economic Review*, Vol. 60 (May 1970): 231-236.
[36] F. Cardoso and E. Faletto, *Dependency and Development in Latin America*. Berkeley: University of California Press, 1979.
[37] Immanuel Wallerstein, *The Capitalist World Economy*. Cambridge: Cambridge University Press, 1979.
[38] Immanuel Wallerstein, "Ecology and Capitalist Costs of Production: No Exit," in *Ecology and the World-System*, edited by Walter L Goldfrank, David Goodman, and Andrew Szasz, 3-11. Westport, CT and London: Greenwood Press, 1999, 8.
[39] Wallerstein (1999), 9.
[40] Quoted in Guha, 126.
[41] Michael E. Zimmerman, ed., *Environmental Philosophy*, 3rd edition. Upper Saddle River, NJ: Prentice Hall, 2001, 3-5.
[42] For the most prominent critique of this perspective, see Ramachandra Guha, "Radical American Environmentalism and Wilderness Preservation: A Third World Critique," *Environmental Ethics*, Vol. 11, No. 1 (1989): 71-83.
[43] Zimmerman, 5.
[44] For a different perspective on critical theories of the environment, see Mathew Humphrey, "Reassessing Ecology and Political Theory," *Environmental Politics*, Vol. 10, No. 1 (Spring 2001), 1-6; and also Karin Backstrand, "Scientisation vs. Civic Expertise in Environmental Governance: Eco-feminist, Eco-modern and Post-modern Response," *Environmental Politics*, Vol. 13, No. 4 (Winter 2004), 695-714.
[45] Carolyn Merchant, *The Death of Nature: Women, Ecology, and the Scientific Revolution*, San Franciso: Harper and Row, 1980; and Merchant, "Gender and Environmental History," *The Journal of American History*, Vol. 76, No. 4 (March 1990), 1117-1121.
[46] Quoted in Merchant (1990), 1117.
[47] Enrique Cienfuegos and Laura Carlsen, "Human Rights, Ecology, and Economic Integration: the Peasant Ecologists of Guerrero," in *Confronting Globalization: Economic Integration and Popular Resistance in Mexico*, edited by Timothy A. Wise, Hilda Salazar, and Laura Carlsen. Bloomfield, CT: Kumarian Press, 2003, 43-64.
[48] Guha (2000), 87.
[49] Eckardt C. Beck, "The Love Canal Tragedy," *The EPA Journal* (January 1979), http://www.epa.gov/history/topics/lovecanal/01.htm; see also Thomas H. Fletcher, *From Love Canal to Environmental Justice: The Politics of Hazardous Waste on the Canada-U.S. Border*. Toronto, Ontario: Broadview Press, 2003.
[50] The groundbreaking empirical study was, United Church of Christ, Commission for Racial Justice, *Toxic Wastes and Race in the United States*. New York: United Church of Christ, 1987. Important academic discussions include, K.S. Schrader-Frechette, *Environmental Justice: Creating Democracy, Reclaiming Democracy*. Oxford and New York: Oxford University Press, 2002; David N. Pellow, *Garbage Wars: the Struggle for Environmental Justice in Chicago*. Cambridge, MA and London: the MIT Press, 2002; and Adam S. Weinberg, "The

Environmental Justice Debate: A Commentary on Methodological Issues and Practical Concerns, *Sociological Forum*, Vol. 13, No. 1 (March 1998): 25-32.
[51] Kristin Schrader-Frechette, *Environmental Justice: Creating Equity, Reclaiming Democracy*. Oxford and New York: Oxford University Press, 2002, 163-184.
[52] Ronald Inglehart, "Value Systems: The Subjective Aspect of Politics and Economics," in *Modernization and Postmodernization*. Princeton: Princeton University Press, 1997, 7-50.
[53] Drawn from Lester Milbrath, 170.
[54] It has become a significant political fact in the United States and other high income industrial countries that organized labor and the management of industrial corporations – traditionally on opposite sides of most social and economic issues – frequently unite against environmental policies that appear to erode industry profitability and employment, and contribute to deindustrialization.
[55] Peritone, "Environmental Attitudes of Indian Elites," in *Asian Survey*, Vol. XXXIII, no. 8 (August 1993), 804-18.
[56] Hsiao, Lai, Liu, Magno, Edles, and So, "Culture and Asian Styles of Environmental Movements," in Lee and So, eds., *Asia's Environmental Movements*, 210-29
[57] Inglehart, 343.
[58] Inglehart, 131.
[59] Inglehart, 242.
[60] See, for example, Guojia huanjing baohu zongju (State Environmental Protection Administration), "Quanguo gongzhong huanjing yishi diaocha baogao (Zhaiyao)" (A Survey on the Nation's Public Environmental Consciousness [Summary]. Huanjing jiaoyu (Environmental Education), No. 4 (1999) 25-27; Yuan Fang, "Zhongguo shimin de huanjing yishi diaocha: Beijing he Shanghai," (A Survey of Chinese Residents' Environmental Consciousness in Beijing and Shanghai). In Xi Xiaolin and Xu Qinghua, eds., Zhongguo gongzhong huanjing yishi diaocha *(A Survey of China's Public Environmental Consciousness)* Beijing: Zhongguo huanjing kexue chubanshe (China Environmental Science Press), 1999, 109-130; Stockholm Environmental Institute, *Making Green Development a Choice: China Human Development Report*. Oxford: Oxford University Press, 2002. For a review of the literature in Chinese from 1990 to 2004, see Yok-shiu F. Lee, "Public Environmental Consciousness in China: Early Empirical Evidence," in Kristen A. Day, ed., *China's Environment and the Challenge of Sustainable Development*. Armonk, NY: M. E. Sharpe, 2005, 60-65.
[61] Lee, 2005, 42, 50, 55.
[62] Ibid, 44.
[63] Ibid. 52.
[64] Ibid. 56.
[65] The questions with percentage disagreeing were: "Plants and animals exist primarily to be used by humans (25.8); mankind was created to rule over the rest of nature (23.1); and humans have the right to modify the natural environment to suit their needs (19.3). Carlos Wing Hung Lo and Sai Wing Leung, "Environmental Agency and Public Opinion in Guangzhou: The Limits of a Popular Approach to Environmental Governance," *China Quarterly*, Vol. 163 (September 2000), 686.
[66] Ronald Inglehart has done path breaking research on postmaterialist values, the root of the NEP. For a summary of Inglehart's findings on postmaterialism, see his *Modernization and Postmodernization: Cultural, Economic, and Political Change in 43 Societies*. Princeton, NJ: Princeton University Press, 1997. See also R. E. Dunlap and K. D. van Liere, "The 'New Environmental Paradigm'," *Journal of Environmental Education*, Vol. 9 (1978), 10, and J. C. Pierce, N. P. Lovrich, T. Tsurutani, and T. Abe ,"Culture, Politics, and Mass Publics: Traditional and Modern Supporters of the New Environmental Paradigm in Japan and the United States," *Journal of Politics*, Vol. 49, No. 1 (1987), 54-79.

[67] Ronald Inglehart, "Public Support for Environmental Protection: The Impact of Objective Problems and Subjective Values in 43 Societies," *PS: Political Science and Politics* (March 1995), pp. 57-71.
[68] Wallerstein (1999), 7-8.
[69] Ibid., 5.
[70] Kevin P. Gallagher, *Free Trade and the Environment: Mexico, NAFTA, and Beyond*, Stanford, CA: Stanford University Press, 2004, 5-6; Gene M. Grossman and Alan B. Krueger, "Economic Growth and the Environment," *The Quarterly Journal of Economics* vol. 110, no. 2 (May 1995): 877-908.
[71] Arthur P.J. Mol, "Ecological Modernization and the Global Economy," *Global Environmental Politics*, Vol. 2, No. 2 (May 2002): 94.
[72] Arthur P.J. Mol, *Globalization and Environmental Reform: the Ecological Modernization of the Global Economy*. Cambridge, MA: the MIT Press, 2001.
[73] World Commission on Environment and Development, *Our Common Future*. New York: Oxford University Press, 1987, 8.
[74] Daniel A. Mazmanian and Michael E. Kraft, eds., *Toward Sustainable Communities: Transition and Transformations in Environmental Policy*. Cambridge, MA: MIT Press, 1999, 18.
[75] See Hans Th. A. Bressers and Walter Rosenbaum, eds., *Achieving Sustainable Development: The Challenge of Governance Across Social Scales*. Westport, Ct: Praeger, 2003, 48.
[76] William M. Lafferty and James Meadowcroft, "Introduction," *in Implementing Sustainable Development: Strategies and Initiatives in High Consumption Societies*. Oxford and New York: Oxford University Press, 2000, 12.
[77] "Concluding Perspectives," 457.
[78] Najam, Adil, "The View for the South: Developing Countries in Global Environmental Politics, in Global Environmental Politics, in *The Global Environment: Institutions, Law, and Policy*, edited by Pamela S. Axelrod, David Leonard Downies, and Norman J. Vig. Washington, D.C.: CQ Press, 2005, 225-43.
[79] For an authoritative introduction to corporatism, see Philippe C. Schmitter and Gerhard Lehmbruch, eds. *Trends Toward Corporatist Intermediation*. Beverly Hills and London: Sage Publications, 1979.
[80] Jonathan Rosenberg, "Mexico: The End of Party Corporatism?" in *Political Parties, Interest Groups, and Democratic Governance*, edited by Clive S. Thomas. Boulder, CO and London: Lynne Reinner., 2001, 247-65.
[81] Olson, Mancur, *The Logic of Collective Action*. Cambridge, MA: Harvard University Press, 1965.
[82] M. Crepaz, "Explaining National Variation of Air Pollution Levels: Political Institutions and Their Impact on Environmental Policy-Making," *Environmental Politics*, Vol. 4, no 3 (August 1995), 391-414.
[83] See Thomas P. Lyon and John W. Maxwell, *Corporate Environmentalism and Public Policy*. Cambridge: Cambridge University Press, 2004, 10.
[84] Lyle Scruggs, "Institutions and Environmental Performance in Seventeen Western Democracies," *British Journal of Political Science*, Vol. 29 (1999), 1-31.
[85] See, for example, Gordon White, Jude Howell and Shang Xiaoyuan, *In Search of Civil Society: Market Reform and Social Change in Contemporary China*. Oxford: Clarendon, 1996.
[86] Cited in Jonathan Unger, "'Bridges': Private Business, the Chinese Government, and the Rise of New Associations," *China Quarterly*, No. 147 (September 1996), 818.
[87] See Greg O'Leary, *Adjusting to Capitalism: Chinese Workers and the State*. Armonk, NY: M.E. Sharpe, 1998, and Hong Ng Sek and Malcolm Warner, *China's Trade Unions and Management*. New York: St. Martins Press, 1998.
[88] See Richard Baum and A. Shevchenko, "The 'State of the State'," in Merle Goldman and Roderick MacFarquhar, eds., *The Paradox of China's Post-Mao Reforms*. Cambridge, MA: Harvard University Press, 1999.

[89] See Chen Feng, "Between the State and Labour: The Conflict of Chinese Trade Unions' Double Identity in Market Reform," *China Quarterly*, No. 176 (December 2003), 1006-38.
[90] See Yep Ray, "The Limitations of Corporatism for Understanding Reforming China: An Empirical Analysis in a Rural County," *Journal of Contemporary China*, Vol. 9, No. 25 (November 2000), 547-66.
[91] Tony Saich, *Governance and Politics in China*. New York: Palgrave, 2001, 209-10; see also Tony Saich, "Negotiating the State: The Development of Social Organizations in China," *China Quarterly*, No. 161 (March 2000).

CHAPTER 3. POLITICAL PROCESSES AND ORGANIZATIONS

1. MOVEMENTS, GROUPS AND PARTIES

This chapter asks large questions about the diversity of groups, organizations, and movements throughout the world pursuing improvement of the environment. It attempts to crystallize the values and attitudes people hold toward environmental issues, discussed in chapter 2, by describing their expression in concrete forms: environmental interest groups, environmental movements, and green political parties. In doing so it moves the discussion one step closer to governments that have the power to act (or decide to not act) on environmental problems.

The chapter begins with a discussion of environmental interest groups, considering their origin in western nations, the types of environmental non-governmental organizations (ENGOs) found today, and important kinds of variations found cross-nationally. Then we turn to the larger environmental movements of several nations, and explore their connection to ecological resistance campaigns. Next we examine the role that organized environmental perspectives play, as political parties, mostly in western nations—exploring their origin, nature, and electoral fortunes. At this point we also consider how media have portrayed environmental groups and movements. The chapter concludes with a discussion of the relationship between the process of democratization and political mobilization on environmental issues, which expresses the broader relevance of environmental politics to changes in the modern state system.

1.1 Environmental Interest Groups

1.1.1. Origins

In the previous chapter, we examined the constellation of values and attitudes of peoples in different nation-states. Here we examine how these views are aggregated by ENGOs, which function as interest groups in the state. We shall emphasize throughout the different incentives (or disin-centives) nations provide for the organization of ENGOs; and we observe the ways in which state structures shape and are shaped by the activities of ENGOs.

The first environmental organizations were formed in western nations, and in response to threats to species and ecosystems from the industrial revolution. Britain hosted the first environmental association. It formed in London in 1863 in response to killer fogs, which enveloped the city and caused hundreds of deaths. The association pressured the London City Council and eventually the Parliament for laws limiting factory emission of pollutants.[1]

The Sierra Club, formed in 1892, was the first North American environmental organization. Its charter objective was to preserve the Sierra Nevada range in the western United States, as well as western animal and plant species threatened by encroachments of human population and extensive economic development.

These early ENGOs began as local responses to human and economic development pressures. Initially, they were grassroots organizations, forming in societies that placed few obstacles in the path of citizen initiatives and organizations. Both Britain and the United States have histories as pluralist, not corporatist societies, and they permit the free association of citizens to achieve policy goals. Early membership of ENGOs was composed of those directly affected by environmental degradation. Their objectives were to preserve ecosystems and threatened species. For the most part, at that time, they did not seek a halt to industrialization, but sought to mitigate its impact on valued areas and species.

The development and growth of these organizations corresponded with national conservation movements in both Britain and the United States. For example, the presidential administration of Theodore Roosevelt at the turn of the twentieth century sought to make Americans aware of the effects of rapid industrial development and waste on natural resources.

1.1.2. Expansion

It was not until years into the post-World War II era, however, that ENGOs significantly expanded in number and membership. Increased evidence of environmental degradation, such as the publication of Rachel Carson's *Silent Spring* in 1963, broadened elite and public understanding of the perils of ruthless development. Of perhaps greater importance, a succession of environmental crises aroused public indignation. The first American mobilizing event was the Santa Barbara oil spill of 1969, which coated beautiful sandy beaches outside this upscale California community with oily slime killing thousands of waterfowl, fish, and other marine species. This led directly to the first Earth Day event on April 15, 1970, which brought about a revolution in environmental organizing.

Most of the current ENGOs in western nations either developed after major national environmental crises, or their formation was influenced by a process of horizontal diffusion of ideas from the "pioneering" nations environmentally, such as the United States and Great Britain. The process was iterative, in that new organizations sought changes in the law, and these changes created incentives for further organization of ENGOs and new public participation requirements in environmental laws.

In the early twenty-first century, most western nations have a full panoply of environmental organizations. Perhaps the largest number of organizations is found in the United States, and they claim the largest membership as well as the healthiest budgets, although these vary by perception of threats presented in different national political coalitions.

1.1.3. Types of ENGOs

There is immense variety in the kinds of ENGOs found within the world's nation-states. In this section we analyze them by six main criteria: scope, membership, leadership, purpose, government connection, and orientation to the state.

Scope. ENGOs range in territorial scope from local to global. As mentioned, local ENGOs are more likely to be grass roots groups than hierarchically ordered, top-down associations. Most formed in reaction to pollution of air or land, including toxic contamination of water, in neighborhoods. As such, they are good examples of NIMBY (not-in-my-backyard) protest organizations, which focus on local developments to the exclusion of national or global patterns of environmental degradation. The largest number of ENGOs, however, are national environmental associations, for example the American Sierra Club and Wilderness Society. Their energies are directed to environmental problems and events within the national boundaries, although they may observe global environmental issues affecting that nation-state. Finally, some ENGOs are truly global (and for this reason are called INGOs, or international NGOs), and they have outposts in a large number of nation-states. The best (and perhaps the most controversial) example of a global ENGO is Greenpeace, with an international board of directors representing a dozen different nation-states, and offices in more than 70 countries. A second example is the World Wide Fund for Nature (WWF), which sponsors projects in many countries and attempts to empower local communities to enable them to control their environmental futures.[2]

Membership. ENGOs also vary by the type of their membership as well as its size. First, in economically developed countries (EDCs), members of environmental organizations disproportionately come from the middle and upper classes. The exceptions to this generalization are participants in local grass roots ENGOs who are more likely to be lower-middle or lower class in economic terms and from ethnic minority communities.[3] Second, ENGOs are likely to solicit dues-paying members, and thus are in the nature of mass-based organizations, instead of caucus-type organizations with a small number of members. In some smaller states and communities, however, leadership and membership of NGOs and community-based organizations (CBOs) may be virtually the same. Depending on their scope and purpose, they may consist of a small "board" of volunteer activists who both make policy and carry it out. Or, they may consist of one or a few activist/leaders and a volunteer or paid staff person. In such organizations support comes not from a stable membership but from contingent, temporary supporters of particular campaigns or functions, or from grants and contracts with governments, international organizations or larger national and international ENGOs.

Leadership. New ENGOs with small memberships may have just one primary leader, who carries the organization based on her/his prestige and may even defray its expenses. Even large ENGOs with mass memberships, may attract charismatic leaders, as seen in David Brower's leadership of the Sierra Club and then Friends of the Earth. The degree of membership control over leaders is a second variable, and this tends to co-vary with scope of the ENGO. In large, mass-membership organizations, there is likely to be an organizational bureaucracy that may have a great deal to do with setting the goals and activities of the ENGO, and determining its bargaining strategy. In local ENGOs, on the other hand, members may expect to be involved in decisions about tactics as well as strategy. The norm of participatory democracy is more likely to apply, however there are important exceptions based on organizational strategy, the characteristics of leaders, and the national political culture within which the organizations are embedded.

Purpose. ENGOs also vary by the extent of their purposes. Some have specific purposes, whether to save a particular species (such as the Save the Whales Foundation), to protect and promote the wilderness (such as the Wilderness Society), or to preserve particular types of ecosystems (such as Oceana and Wetlands International). Other ENGOs have multiple purposes. For example, Greenpeace has interests in reduction of greenhouse gas emissions, species preservation, GMO regulation, and elimination of toxic waste proliferation. Moreover, there is a correlation between the age of the ENGO and the variety of its purposes. Thus, the Audubon Society began with a focus exclusively on preservation of bird species and conservation of bird sanctuaries, but recently has enlarged its objectives to include broader ecosystems. And the Sierra Club has become concerned with marine as well as terrestrial ecosystems, and with endangered species preservation work.

Governmental Connections. In the EDCs, most ENGOs are private, non-profit associations, and have no formal linkage to government agencies, whether the state environmental protection agency or another department. The exception is state funding of ENGOs in several European countries. Thus,

Norway has a large number of QUANGOs, or quasi-NGOs, which are financed by the state but have autonomous agendas. In the lesser developed countries (LDCs) on the other hand, a number of ENGOs may have been established by governmental agencies, as one means of assisting in the implementation of environmental policy. For example, scholars have identified nearly 350 ENGOs in the People's Republic of China. Most of these associations, however, are properly called government-organized NGOs, or GONGOs. The State Environmental Protection Administration, China's foremost environmental agency, has a dozen such GONGOs to assist it in its work. Interestingly, there is also a tendency toward GONGOs in small, democratic states where populations may be too small and poor to establish or sustain true NGOs. For example, in Dominica the National Association of NGOs (NANGO) channels government support and recognition to several small environmental and social service organizations, and the Fisheries Division of the Ministry of Agriculture helped organize the non-governmental Local Area Management Authority for a marine protected area.

Orientation toward System. The final criterion refers to whether the ENGO works within the existing national system and has a pragmatic orientation, or attempts to change significantly the status quo organization of power, and can be called "radical." For obvious reasons, ENGOs in authoritarian countries operate within the system, but their very existence may challenge the state, as we discuss below in the section on democratization. Similarly, in authoritarian systems, the state may require that ENGOs meet arduous registration requirements in order to operate. In 1998, China's Ministry of Civil Affairs required that NGOs be sponsored by a state agency; membership in the group was limited to 50 and it had to demonstrate fiscal responsibility. A group was required to seek separate registration in each place it operated, which effectively prohibited the growth of national associations not directly part of a national bureaucratic agency. Also, only one social association could register in the same area for a specific activity, such as environmental protection. As a result, "NGOs tread carefully, avoiding strong criticism of governmental environmental protection failures."[4]

In North America and Europe, there is a clear distinction between "mainstream" and "radical" environmental organizations. In the United States, mainstream ENGOs are affiliated with the "Group of Ten," and collaborate in planning on national issues. The Group includes the Wilderness Society, Friends of the Earth, the Sierra Club, the National Audubon Society, the Environmental Defense Fund, the National Wildlife Federation, the Izaak Walton League, the National Parks and Conservation Association, the National Resources Defense Council, and the Environmental Policy Institute. Most are vast membership associations[5] with large professional staffs which track environmental legislation in Congress and in administrative agencies and lobby the Congress on environmental issues.[6]

A number of organizations, while not explicitly calling for violence to accomplish their objectives, nevertheless favor direct action, including civil disobedience, obstruction of economic development threatening environmental values, and nonviolent demonstrations. Greenpeace and the Sea Shepherd Society fall into this category with respect to their attempts to disable commercial fishing vessels threatening dolphins. Some smaller radical ENGOs, such as the Animal Liberation Front, Earth Liberation Front, and Earth First! seem to have condoned forms of violence, such as "monkey-wrenching" and spiking trees planned for harvesting in old growth forests.[7] Ecological resistance groups provide many examples of direct action.

1.1.4. National Variation in ENGOs

The preceding indicates chief factors influencing the variation of ENGOs in different countries. The economically developed countries are likely to have a broad span of environmental organizations, many of which will have quite large memberships, drawn from the middle and upper reaches of society. Most of their ENGOs will be private, non-profit associations. They are autonomous agents in civil society without relations of dependency to government agencies. While they are likely to have both specific and general-purpose ENGOs, as the associations age, in order to survive, they will add additional functions. Finally, because these states are likely to have capable governments, most of the ENGOs will work within the system instead of in opposition to it.

In LDCs, on the other hand, one tends to find fewer national-level ENGOs, and relatively more at the grassroots and global levels. Their membership is less likely to represent the middle classes alone; their grassroots organizations may represent large numbers of poor and minority individuals. They are more likely to have specific purposes, related to direct threats to the environments of villages and towns, than general purpose associations. LDCs are more likely to have GONGOs, as one means to help resource-strapped governments administer environmental policy. Finally, the "implementation deficits" of LDCs are incentives for the development of ecological resistance movements, which we shall discuss below.

The EDCs are all liberal democracies (although this does not include all "high income countries"); however, LDCs are divided between democratic and authoritarian state systems. Above we noted that authoritarian states prohibit formation of radical ENGOs, although they may not have the capacity to prevent them entirely. Their preferred strategy is to play an active role in establishing GONGOs, which are less likely to be found in democratic state systems. Therefore, the autonomy of legal ENGOs in authoritarian states is limited.

National variations also derive from differences in political culture, state structures, and state capacity (frequently a function of levels of economic development). Along those lines, we see variations not only between democratic and authoritarian systems, but among democratic systems.

In the small island developing states (SIDS) of the Eastern Caribbean, parliamentary democracy and a general respect for civil liberties allows the formation of autonomous ENGOs. But small population size, poverty and economic dependency create particular challenges. Political parties, which are the key institutional actors in the formation of parliamentary governments, have small popular bases and tend toward personalistic politics and charismatic leadership. The institutional capacity of both governments and NGOs is limited by a lack of financial and human resources. Trained scientific and administrative personnel must go abroad for training and are often lost to "brain drain." External aid and support, therefore, become important contributors to organizational efficacy. ENGOs that can bring that support home may be influential.

The effectiveness of the Dominica Conservation Association (the country's only exclusively environmental NGO), for example, rested largely on the efforts and reputation of its director, Atherton Martin, a prominent figure in national politics, as well as the tourism and agriculture industries (discussed further in chapter 6). However, his political opponents saw his ability to attract external funding and affect public opinion in Dominica and internationally as related to his personal political ambitions.[8] After winning the Goldman Prize in 1998 (a large cash award given annually by a private foundation to grass roots "environmental heroes" in six world regions)[9] he joined government in 2002 as Minister of Agriculture, Planning and the Environment. But he resigned after only six months to protest the government's support of Japan in the International Whaling Commission.[10]

In Grenada, small, multi-purpose ENGOs have made important contributions. For example, during the late 1990s the leader of a small wildlife preservation society was able to postpone a solid waste disposal project by testifying at a public hearing convened by the World Bank (the project's lead funding agency) that the land designated for the facility was habitat for the endangered national bird, the Grenada Dove. A local environmental consulting firm (virtually a one-woman operation) was then hired by the Government of Grenada to find a new location for the facility. The consultant voluntarily took on the additional functions of an ENGO by including a public information campaign and opinion survey on environmentally responsible waste disposal to fulfill the World Bank requirement for public consultation.[11]

A few ENGOs that focus their activities on all or part of the region have also been important actors in the environmental politics of Caribbean SIDS. The Caribbean Natural Resources Institute (CANARI) based in Trinidad and St. Lucia, advocates grassroots participation, conducts research on participatory management practices, and facilitates the development and formation of locally based management plans and institutions.

> The central aim of the programme is to foster the development and adoption of policies that support increased participation and collaboration in managing natural resources. The starting point is research and analysis, in order to understand the institutional arrangements, skills, technologies, support mechanisms, and processes of policy formulation and reform required for the adoption of participatory approaches. The knowledge gained from this analysis is used to promote policies, through a systematic process of advocacy. The programme of the Institute is structured in such a way that the links between its three elements, namely applied research, analysis, and advocacy, are reinforced.[12]

CANARI's interventions—at the invitation of governments, international organizations, aid agencies and other NGOs—include providing personnel to facilitate policy development and institutional design, and mediate environmental disputes. In addition, CANARI conducts research and disseminates information on participatory resource management. As such CANARI has become an important adjunct to government, NGO and community-based organization (CBO) capacity in the region, as well as an effective advocate for participatory and deliberative democracy in British-style parliamentary systems that otherwise tend to be highly centralized in their decision-making processes.

Grassroots ENGOs are difficult to form and maintain in LDCs, even those bordering rich EDCs, and they require strategic alliances with national and international groups. A notable example is the developing effort to protect mangrove forests and wetlands along the Mexican coast. Mangroves themselves are valuable species for biodiversity protection. The coastal lagoons they inhabit also are home to endemic plant, insect, and marine species, such as tiger sharks. Mexican coasts are under intensive pressure from developers of resort hotels, upscale housing, golf courses, and industrial zones for marine terminals. Yielding to pressures from developers with political connections, the Mexican federal Natural Resources and Environment Ministry secretary made a small change in the state's strong mangrove protection law, which permitted cutting mangroves if they were replanted in other areas and opened the door for hundreds of tourist ventures.

The leader of a grassroots effort to protect mangrove forests in the northwest region commended: "Sometimes government is our worst enemy. They don't enforce the law; they hide information."[13] To raise consciousness, this grassroots effort has reached out to internaitonal NGOs such as The Nature Conservancy, Conservation International, WWF, and Global Green Giants Fund, which provide some funding and technical support (local ENGOs are poorly funded and require assistance to negotiate the complex legal environment). Of greatest assistance are scientists, such as those affiliated with the Research Center on Food and Development (CIAD), which manages several protected areas and a regional natural history museum in the Estero del Yugo (a low tropical decisuous forest adjoining a coastal wetland).

In larger, presidential democracies like the United States, the structure of the system as well as the greater capacities of governmental and non-governmental institutions lead to different and more differentiated strategies for ENGOs. For its activities in the U.S., the Sierra Club has developed a complex institutional structure with divisions for environmental education, field trips, publications, green investments, political action, lobbying, and vetting candidates for federal office and judicial appointments. The organization also reflects the federal structure of the United States with chapters in all 50 states. In 1971, the club founded its own, non-profit environmental law firm, the Sierra Legal Defense Fund, to pursue environmental protection through the courts by litigating claims against government and private entities and challenging governments on the implementation and constitutionality of environmental laws. In 1997, the Legal Defense Fund was renamed and reconstituted as Earthjustice with its own board of trustees and separate membership.[14]

1.2 Environmental Movements

Of course there is more to environmentalism in different nation-states than environmental interest groups alone. What seems to explain the large size and robust character of ENGOs in many countries, particularly those in EDCs, is the intimate relationship with a national environmental "movement." The movement is all-encompassing and multi-phasic; it is the sauce in which ideas about environmental change are cooked, before they become policy suggestions and products. We explore the various meanings of the movement concept, apply it to ENGOs in some EDCs, and then explore whether ecological resistance movements of LDCs fit within the concept or not.

1.2.1. The "Movement" Concept

A developing literature in social movement theory helps us understand the environmental movement. This literature[15] defines a "movement" in three parts. First, it represents the mobilization and organization of large numbers of people to pursue a common cause. The cause may be improvement of the environment, enhancement of working conditions for labor, upgrading the status of women, or spreading civil rights to all citizens, with a specific focus on racial and ethnic minorities. However, in the life-cycle interpretation of movements,[16] the mobilization is not enduring and typically peaks within a generation (20-30 years); in the process of development it prompts a backlash, which may bring about its demise.

The second part of the definition is the community of believers created by the mobilization process. Finally, the movement is a stage (the second) in a continuum of at least four stages: (1) Initially, people respond spontaneously and in an unstructured fashion to public problems. This stage is characterized by fads, crowds, and anomic action. (2) The social movement as the second stage still has aspects of spontaneity and fluidity, but there is some organizational structure. (3) At the third stage, interest groups, spawned during the movement, become institutionalized. They participate in policy-making but do so outside the formal authority structure. (4) At the fourth stage, movement advocates have become elected or appointed officials of the state. Their causes are part of the agenda adopted as public policy.[17]

Other, structural characteristics of the social movement are its segmentary nature, polycentric leadership, and reticulation. The movement is segmentary because it is composed of many diverse groups; some grow, some expire while pursuing movement objectives. The movement is polycentric in that it has several leaders and many centers of influence. Finally, the reticulation of the movement refers to the fact that it is a loose, integrated network with multiple linkages and overlapping membership.[18]

Effectively, the movement ceases to exist as a movement when it has become institutionalized. This means that when the leaders of the movement and its goals have been incorporated into formal instrumentalities of the state, such as political parties, then it has become part of conventional politics.

1.2.2. Examples of Environmental Movements

By this elaborate definition, states with established Green parties, which participate not only in electoral politics but also in coalition governments, have progressed beyond environmentalism as a movement. There is at least one nation-state, to wit Germany, which has attained this stage of movement development and transformation. At the other end of the continuum, some authoritarian states, such as North Korea, prohibit formation of civil organizations, and thus response to environmental problems is entirely spontaneous, unstructured and usually short.

The United States has one of the oldest environmental movements in the world. Between the establishment of new ENGOs in the 1960s and Earth

Day in 1970, nearly 1 million Americans had become members of national groups, and as noted above, these groups in the 1970s and 1980s began to work together as part of a national alliance or coalition. Members of the environmental movement were diverse ideologically, with mainstream national groups increasingly opposed by somewhat radical splinter organizations. Also, goals of the movement were diverse with respect to areas of emphasis and national as compared to state and local orientations. Nonetheless, most members of the environmental movement shared values of human-nature interaction, and were in general agreement on basic movement goals and beliefs. The large membership of ENGOs in the United States, strategy of environmental leadership, and developing linkages with leadership nationally and at the state and local levels, gave the movement significant clout, as indicated by the large body of national legislation enacted in the 1970s. This clout also brought on a backlash in the 1980s, which perhaps refreshed the movement instead of causing it to collapse.

A quite different example is seen in China. In this country, environmentalism developed after the state had established an environmental ministry and national environmental laws. As part of the political liberalization of the mid-1990s, the authoritarian regime allowed some groups to form environmental organizations, largely as a means to further the implementation of state policy on environmental issues. The regime also permitted foreign ENGOs to operate because they brought new resources from abroad to address problems of biodiversity conservation and pollution reduction. Development of ENGOs occurred simultaneously with the liberalization of news media, which could report more freely on environmental than social or political problems of the state. By the early twenty-first century, China had entered the stage of organizing large numbers of people into associations with environmental goals, many of which were directly under state control. This clearly retards the autonomous mobilization phase, regarded as essential to the growth of environmental movements.

1.2.3. The Status of Ecological Resistance Movements

Ecological resistance must figure in comparative treatments of environmental movements, but this type of social and political protest does not easily meet the definitions of the movement concept provided above. Ecological resistance movements are more expansive, in a territorial sense. They are reactions to several oppressive stimuli, and thus have a broader set of goals and objectives. Movement participants differ from those found in typical environmental movements, and they employ the tactics of direct action. We examine these differences through a review of several cases presented in Bron Taylor's *Ecological Resistance Movements*.[19]

We noted that environmental movements were one stage of a continuum, and not found in all countries of the world. Particularly, authoritarian states discourage formation of environmental interest groups, and at the other end of the continuum, some states have embedded environmental action and officials into the authority system. We hypothesize, however, that ecological resistance movements have formed (or may form) in virtually all nation-states. They are more universal, because they can originate as spontaneous responses to pressure, and they do not require a specific organizational form in order to exist.

Environmental movements arise in response to environmental problems, but ecological resistance is a response to multiple causes of disenfranchisement, disempowerment, and dissatisfaction. These causes vary depending on a state's level of economic and political development. For instance, most LDCs were subjects of colonialism and imperialism, which weakened or destroyed indigenous authority structures and human-environmental patterns of relationships, creating a situation of dependency. Even upon independence LDCs easily can become disadvantaged by multi-national logging, mining, hydro-electric, and oil/gas corporations. They can be dominated handily by great powers such as the United States. Domestically, local populations may be oppressed by national elites linked to multi-national corporations or great powers; the extension of the market system may disrupt traditional patterns of land use. Local populations also may suffer under continuing but delegitimized hierarchical authorities. Yet ecological resistance also is found in EDCs without a history of external domination or dependency. Nevertheless in such countries, local populations may feel oppressed by policies of internal colonialism and exploitation by national or multi-national corporations.

The diffuse range of oppressive stimuli prompts a broader response. Instead of NIMBYism, the victims' response to particular cases of environmental degradation, ecological resistance expresses an orientation of opposition to social and economic inequality. For this reason, ecological resistance is often called a movement for environmental justice. In most cases and particularly in Africa and many parts of Latin America and Asia, the focus is on the relationship between the people, their culture and livelihoods, and both the land tenure system and the broader environment; it is not a specific response to environmental problems alone.

Fourth, the participants in ecological resistance movements are different from members of environmental organizations in most EDCs. Many more women are involved in such movements, and a number of protest groups are headed by women. It can be argued that women feel ecological crises most acutely because they are primary caregivers for children, spouses and extended family and providers of family subsistence. Also, they are most oppressed by patriarchal authority structures and subjects of discrimination nationally and locally.

Racial and ethnic minorities also are disproportionately represented in ecological resistance movements. They are least likely to be integrated into national economic and political life, often live at the periphery of the power centers (which are urban), and usually lack instruments through which to communicate their needs. Also, participants represent a larger number of lower socio-economic groups, to which the political system customarily responds poorly if at all. This combination of elements does not lend itself to typical interest group mobilization and organization. It may, however, be assisted by external agencies, such as the role played by the Catholic Church in ecological resistance movements of Central and South America and in the Philippines.

The tactics of ecological resistance movements encompass direct action including activities that are illegal. For example, in India, the Chipko (to hug) ecological resistance movement developed in opposition to logging and deforestation. In Africa, the poaching and illegal trafficking of endangered and protected species are acts of resistance that can be interpreted as means to support livelihoods and confirm life systems. In many parts of the developing world ecological resistance has arisen as an element of indigenous peoples' movements.

The distinction between environmental and ecological resistance movements is academic, and in reality it is not always easy to discern. But separating the two concepts does capture some important differences in the environmental politics of democratic EDCs and LDCs, both authoritarian and democratic. For one, authoritarian states have a high propensity for radicalizing peaceful movements. Groups originally organized to carry out conventional and ostensibly legal forms of protest may turn violent in the face of repression. States that are only formally democratic or still in the early stages of transition to democracy will retain some authoritarian tendencies and respond coercively to social protest movements. And even established but underdeveloped democracies may lack the personnel and institutions to respond effectively to peaceful challenges to authority and property. In addition, governments (both national and sub-national) in LDCs may lack the capacity or will to protect the rights and even the lives of environmental protestors who run afoul of powerful private interests.

1.2.4. Three Cases of Environmental Resistance

The Ogoni people, despite the location of their homeland in the oil-rich Niger River Delta, are among the poorest ethnic groups in Nigeria. Since the 1960s they have sought greater autonomy from the national government, more economic benefits from the oil deposits beneath their lands, and redress for the severe degradation of agricultural land, fisheries and water sources caused by frequent oil spills, wellhead blowouts and gas flaring. In 1990, the

Movement for the Survival of the Ogoni People (MOSOP) presented the military government of Nigeria with an Ogoni Bill of Rights.

> The Bill of Rights . . . asserted that . . . autonomy should guarantee the right to control their political affairs and to the control and use [of] a fair share of the economic resources derived from Ogoniland, the protection, use and development of Ogoni local languages, and the protection of their oil-producing environment from further degradation.[20]

Ken Saro-wiwa, an internationally known playwright and leader of MOSOP, took the position that the solution to the environmental degradation, poverty and the repression suffered by the Ogoni under a series of civilian and military governments lay in restructuring federal Nigeria as a confederation of semi-autonomous, ethnically defined regions, with the bulk of the revenues from each controlled by regional authorities.[21]

In May 1994, a public rally turned into a riot and four Ogonis were killed. The military government blamed Saro-wiwa and eight other MOSOP leaders for the killings. Outside observers agree that the guilty verdicts and death sentences handed down by the federal court were not supported by the evidence and would have been overturned given the opportunity of an appeal. Nevertheless, on November 10, 1995 the nine were hanged.[22] In his closing remarks to the court, Saro-wiwa condemned not only the Nigerian military but the major oil multi-national operating in Ogoniland.

> I repeat that we all stand before history. I and my colleagues are not the only ones on trial. Shell is here on trial and it is as well that it is represented by counsel said to be holding a watching brief. The Company has, indeed, ducked this particular trial, but its day will surely come and the lessons learnt here may prove useful to it for there is no doubt in my mind that the ecological war that the Company has waged in the Delta will be called to question sooner than later and the crimes of that war be duly punished. The crime of the Company's dirty wars against the Ogoni people will also be punished.[23]

The rubber tappers of the Brazilian Amazonian state of Acre have long been economically marginal.[24] Originally brought to the region in conditions of virtual debt peonage during the rubber boom of the late nineteenth century, they have eked out an existence through a combination of subsistence farming, and selling rubber and brazil nuts. In the 1970s, the military government fueled a rush for Amazonian land by making public lands available for cattle grazing and speculation, and subsidizing developers with road building projects. The resulting closure and clearing of the lands jeopardized the existence of the rubber tappers.

In 1974, the year that the military government officially began the process of democratization known as *abertura* (opening), the rubber tappers began to organize, with the assistance of the Catholic Church and the National Confederation of Agricultural Workers (CONTAG). The 1970s was a growth period for Brazilian civil society, as a myriad of new popularly based groups advocating a range of civil liberties and social justice issues arose to take advantage of a more open environment and accelerate the process of democratization. The claims of the tappers to their lands were rarely documented making it difficult for them to pursue their interests through the legal system. The principal tactic of the tappers then was to impede the work crews sent by the new owners to clear the forest. The *empate* or standoff, was meant to be a non-violent confrontation. But the organized tappers' actions were frequently met with violence by police, military, and private paramilitary units employed by ranchers and speculators.

Elected civilian governments continued to encourage development in the Amazon and extended roads further into the rainforest (often with money borrowed from foreign sources and international lending institutions such as the World Bank). State and local governments (Brazil is a federal system) were either supportive of or incapable of controlling attacks on the tappers as they organized to preserve the forests that provide their livelihood. However, a reform of the national law on political parties in 1979 allowed the free formation of parties for the first time since 1964. That year the Workers Party (PT) was formed in Acre. The PT organized in alliance with CONTAG, and supported the formation of the National Council of Rubber Tappers.

The ranchers responded with violence, assassinating the president of a rural workers' union locals. The rubber tappers, under the leadership of Chico Mendes broadened their tactics, developing educational and economic strategies for their membership, and forming an alliance with international environmental activists by framing the rubber tappers' cause in terms of preservation of the Amazonian rainforest. On December 22, 1988 Chico Mendes was murdered. In 1990, the federal government declared the first "extractive reserves" in the Brazilian Amazon. In 2002, Luiz Inácio Lula da Silva of the PT was elected president.

Mendes and the rubber tappers have been hailed as exemplars of grassroots environmental activism and effective global networking for challenging authoritarianism and environmentally unsustainable development. It is impossible to say with certainty if the rubber tappers contributed significantly to democracy in Brazil, but their story offers an illustration of the ways in which democratization can change the relationship between environmental movements and the state.

Controversy over the social, cultural and environmental impact of big dam projects has catalyzed important developments in the environmental movement in India. Historically, big dam projects have played a key role in Indian development strategy.

> Independent India's first generation of leaders considered hydroelectric projects 'temples' of modern India. So strong was the pro-dam bias that the interests of project-affected people were not regarded as central to the planning of these projects. Displaced families were given some monetary compensation, and then they were forgotten.[25]

By 1997, an estimated 20 to 50 million people had been forcibly evacuated by the Indian government for dam construction. Depending on the project, between 55 and 98 percent of the evacuees were members of tribal groups.[26] Opponents to the dam projects not only decry the treatment of the evacuees and the damage done to their culture and social structures, but claim that economic benefits of the dam projects (increased electrical generation, flood control and irrigation) are significantly outweighed by the environmental costs.

Among the largest of these Indian dam projects is the Narmada River Development Project, which originally included the construction of thirty large dams and dozens of smaller ones. One of the larger dams, the Sardar Sarovar Dam in Gujurat state, has been the focal point of the protests and litigation by the Narmada Bachao Andolan (NBA), a coalition of ENGOs and CBOs. NBA includes groups of affected tribal populations, leading Indian intellectuals and environmental activists, and a network of international supporters including the U.S.-based International Rivers Network. The World Bank was the original target of anti-dam activities and responded with major revisions to its lending policies, including requirements for thorough environmental impact assessment and resettlement policies, pledging that "no affected person should be made worse off by a bank funded project."[27] When the Indian government failed to meet the bank's conditions the bank suspended funding in 1992.

Linking environmental and human rights issues enabled the NBA to pressure the World Bank to withdraw but did not stop the dam project. The Indian government remains determined to complete the Sardar Sarovar Dam. NBA has organized well-publicized peaceful demonstrations to impede dam construction, and participants have been arrested for acts of Ghandian civil disobedience. Without success, they have tried to stop dam construction by challenging the constitutionality of the project in the Indian Supreme Court. And in 1999, internationally known author Arundhati Roy was released from custody by the Supreme Court but chastised for publishing account of the Court's decision on the dam that "[scandalized] the court" and "[presented] a one-sided and distorted picture of the proceedings."[28] Since then, NBA has continued its protests, litigation and public information campaigns and scored some successes in exposing official corruption and having some restrictions and mitigating measures added to the project.[29]

Significantly, in all three cases local actors were motivated by threats to their livelihoods, if not their very existence. Culture, ethnicity and conflicting concepts of economic development figured to varying degrees in each story; and local and national governments were weak, inaccessible or hostile when initially approached. Therefore, international prominence became important for all three movements as their primary targets shifted between the national and global levels—oil multi-nationals in the Nigerian case, the World Bank in the Indian and Brazilian cases, and global networks of environmentalists in all three cases. These characteristics highlight important differences between EDCs and LDCs that have a direct impact on environmental movements. Globalization challenges the sovereignty of poor, dependent states more than wealthy states. This is a good news/bad news situation for environmental movements. Locally based movements, if they network effectively, can pressure global actors who in turn can pressure (or at least withdraw support from) national governments. But as the continued pace of deforestation in the Amazon attests, democratization is likely to have limited near-term benefits for the environment as long as national governments lack the capacity to overcome powerful competing interests and attack the sources of environmental problems.

In summary, the ecological resistance movements of many nation-states represent an attempt to redefine the status of the powerless, not only with respect to their immediate environment, but also regarding their membership and participation in the national community. Typically, they link several issues. Environmental degradation is seen as part of a broader and deeper pattern of repression that has social, cultural and economic consequences, and raises questions about the legitimacy of the state.

1.3 Green Parties

Green parties symbolize the institutionalization of environmental movement protest and ENGO activity. Where they have met with electoral success, they are examples of effective interest articulation, bringing the political demands of individuals and groups for environmental protection into more comprehensive policy programs. For example, environmentalists' demands for improved water quality are not the only demands on the state; they are balanced by the interests of businesses, and public preferences for lower taxes, among other competing demands. An environmental program becomes politically significant when it is backed up by substantial political resources, such as popular votes, commitments of campaign funds, seats in the legislature, positions of executive influence, media access, or even armed force.

Political parties are important in interest aggregation in democratic and in many non-democratic nations. Each party (or its candidates) stands for a set of policies and tries to build a coalition of support for this program. In a democratic system two or more parties compete to gain support for their alternative policy programs. In authoritarian systems, a single party or institution may try to mobilize citizens' support for its policies. In both systems interest aggregation may take place within a political party; for example, party leaders hear the demands of different groups and create policy alternatives. In authoritarian systems the process is frequently covert and controlled, and interests often are mobilized to support the government, instead of the government responding to public interests. In either democratic or authoritarian systems political parties seek a direct role in the exercise of political power. That is to say, they attempt to gain direct input into (rather than just influence) state decision-making. In this they are different from interest groups.

Since the 1980s, what was once only an interest group or movement—the environmental movement—has become a factor in election campaigns and party politics. Green parties are found today in most European nations, in North America, and in some Latin American and Asian nations; they have experienced greater success in multi-party than in two-party systems. In this section we examine the origin and nature of Green parties in European countries, the contribution of electoral institutions to their rise and growth, their electoral fortunes, and whether their status in the early twenty-first century constitutes evidence of a "new politics." For comparison, we also include a discussion of one party in an LDC transitioning to democracy.

1.3.1. Origin

Green parties developed first in Great Britain (in 1973) and a few years later in both France and Belgium (1978). By the late 1980s, they were found in 17 countries.[30] In 12 European countries, Green parties have elected representatives to the parliament. Finally, in 5 countries, Greens have entered national governments: France, Germany, Italy, Belgium, and Finland. Most well-known are the German Greens, which joined the Social Democrats in a red-green governing coalition after the 1998 federal elections and stayed in government until the Christian Democrat's victory in the 2005 elections; their de facto leader, Joschka Fischer, took the post of foreign minister.

Most scholars studying the rise of Green parties note that a primary precondition for their emergence is postindustrial social organization, which develops only in rich countries. Two other factors appear correlated with their emergence. First, political opportunities promoted their rise. In the 1980s they grew in those countries where Social Democrats had been regular government participants. Second, centralized corporatist patterns of labor union organization with low strike rates are associated with the strength of left-libertarian parties.[31] Ecological hazards may stimulate the rise of Green parties, as occurred in Sweden after Chernobyl, but there does not appear to be a strong linkage of ecological disasters to the rise of Green parties cross-nationally.

1.3.2. Nature

Green parties have the appearance of being single-issue parties, but in all countries they have developed appeals broader than environmental politics and issues such as nuclear power. In the words of Kitschelt, they present a "new paradigm on ecology." They seek to protect nature because of its intrinsic value, as an object of aesthetic human enjoyment, and for human health, considerations which cannot be valued in economic terms or for reasons of political expedience alone. In addition to environmental protection, they typically strongly support women's emancipation and equality, civil rights and minority protection, a comprehensive welfare state, disarmament, and aid to developing nations.

This issue base suggests that the Greens depict a new face of postmaterialist values. They are unlike the old left parties seeking economic redistribution, and do not seek to replace markets by authoritarian state power. Instead, they search for libertarian institutions to enhance personal and small group autonomy, voluntary associations, and democratic self-governance. Their support base includes younger, educated voters, professionals, and public sector employees.

Even where conditions are ripe, the transition from movement to viable party is not seamless, however. In Germany the Greens made a relatively rapid transition from movement to member of a governing coalition; but they were changed by the process. The party that was officially born in 1980 grew out of the radical youth movements of the so-called Generation of '68 that included violent Maoist and Trotskyite revolutionaries such as the Red Army Faction. The Greens were also steeped in leftist ideology but were pacifists who espoused removing Germany from NATO, raising the price of gasoline dramatically to discourage automobile use, and legalizing "recreational" drugs. They first gained representation in the Bundestag (the lower house of the German parliament) in 1983, and improved their numbers in 1987. Electoral success contributed to a fracture between the "Realos"—political realists who felt that the best way to achieve the party's goals was to strengthen its position in electoral and legislative politics through compromise—and the "Fundis"—fundamentalists who strove for a full transformation of German society and politics. The opportunity to form a coalition with the Socialists in 1998 cemented the ascendancy of the Realos. As members of the governing coalition, the Greens (and Foreign Minister

Fischer) supported sending military missions to Kosovo and Afghanistan, and even compromised on their position on nuclear power. "Over 25 years, the Greens have managed to arrive in the political middle. They not only are the strongest environmental party in the world, but they have tamed their early idealism to match the reality of politics."[32]

1.3.3. Electoral Institutions

Green parties are found in most economically developed countries, but they are prominent political actors only in nation-states with multi-member electoral districts and proportional representation elections. In states having single-member electoral districts with plurality elections, Green parties stand little chance of electoral success. Indeed, no country with a majoritarian electoral system also has a strong left-libertarian party.

The differences among electoral institutions illustrate the problems Greens face. As Giovanni Sartori notes:

> In majoritarian systems the winner takes all; in proportional systems winning is shared and simply requires a sufficient share . . . In majoritarian systems the voter's choice is funneled and ultimately narrowed into one alternative; in proportional systems voters are not forced into concentrating their vote and their range of choice may be quite extensive. On the other hand, majoritarian systems propose individual candidates, persons; proportional systems generally propose party lists.[33]

Sartori's is the most recent careful analysis of the impact of electoral systems on party fortunes. He notes in general terms that majoritarian systems "keep party fragmentation low."[34] At the least, they discourage formation of third parties, such as the Greens. Thus, notwithstanding formation of Green parties in Great Britain, Canada, Australia, and the United States in the 1980s, none has sent representatives to the national legislature or participated in government.[35] In one of the most interesting displays of third party national influence, the American Greens' 2000 presidential candidate, Ralph Nader, gained 3 percent of the popular (but none of the electoral college) vote. Because of the close division of the popular vote in many American states, Democrats called Nader a "spoiler," drawing sufficient votes from Democratic candidate Al Gore to deny him victory in the Electoral College, notwithstanding his 500,000 vote lead over Republican candidate George W. Bush in the popular vote.

There is no question that proportional systems provide the greatest advantage to new parties. If these parties can establish their "relevance," meaning having either a "coalition potential" or a "blackmail potential,"[36] they may join coalition governments as the Greens already have done in five European states. The old argument against proportional representation systems was that they led to increased governmental instability, as no party likely would possess a majority of parliamentary seats. A number of countries, such as Germany, have reduced the number of political parties (and thus tendencies toward instability) by requiring parties to attract a minimum number of votes, in the German case 5 percent, in order to limit this occurrence. It is the case that proportional system elections are more likely to lead to coalition than single-party governments, but as Sartori comments, this "does not *necessarily* lead, then, to quarrelsome and stalemated coalition governments."[37]

1.3.4. Electoral Fortunes of the Greens

At their rise into European electoral politics in the 1980s, the Greens were often characterized as single-issue, "flash-in-the-pan" parties, which would disappear once their core environmental ideas had been adopted by established parties, particularly the socialist ones. However, this has not been the case and the Greens have become a stable presence in European party systems.

Muller-Rommel tracks the performance of European Green parties in a total of 59 national parliamentary elections between 1978 and 2000. He finds that their average electoral results range from 1.5 percent to 7.3 percent. Although this is a low rate of electoral success when compared to the established parties, the Greens did not seek to become large parties. Notwithstanding failures in some recent elections, there also have been increased electoral successes in countries such as Belgium, Finland, and the Netherlands. Thus, Muller-Rommel concludes that "there is no empirical evidence to show a significant overall decline of Green parties' electoral performance."[38]

Moreover, in most of those countries in which Greens have participated in government, they have received a significantly higher percentage of the total vote — from 6 to 14 percent.[39] Of the five countries in which Greens hold office in government — Belgium, Finland, France, Germany, and Luxembourg — and the three additional states in which they are potential members of governing coalitions (Austria, the Netherlands, and Sweden) they have accumulated an average of 7 to 35 parliamentary seats.

The Greens have encountered several limitations to the expansion of their electoral and parliamentary success. Environmental issues have become mainstream in most post-industrial societies, and the Greens are not the only party attempting to attract voters with environmental appeals. Second, as a left-libertarian party, the Greens are less attractive as a coalition partner to center-right parties. Third, the Greens' emphasis on grass roots democracy gives party leaders (who turn over more frequently than leaders of larger parties) less latitude to wield influence within coalition governments and exposes them to the risk of fragmentation when coalition governments take positions opposed by grassroots Greens. This was seen, for example, in the conflict of Greens over military involvement in the Balkans and transport of nuclear waste in Germany. As Poguntke notes:

> [G]overnment incumbency required acceptance of the constraints of domestic and international policy-making even if this meant alienating a considerable portion of movement activists who no longer regarded the Greens as an adequate and trustworthy mouthpiece for their concerns and therefore withdrew their electoral support.[40]

Mexico is a substantially different case from the countries discussed above. Not only is it just beginning a transition from a single-party-dominant, authoritarian system, but it is a presidential rather than parliamentary system. Most national power still resides in the president, who is directly elected, and legislative majorities are not required for the formation of governments. Nevertheless, a series of electoral reforms in Mexico have redounded to the benefit of the *Partido Verde Ecologista Mexicano* (Mexican Green Ecologist Party—PVEM). Although still predominantly a single-member-district plurality system, beginning in the 1980s an increasing number of seats in the National Assembly (the lower house of a bicameral legislature) were rewarded according to the percentage of votes (Proportional Representation—PR) polled by a party's candidates on a nation-wide basis. Originally, meant as a token to depressurize an electorate becoming restive about authoritarian control and electoral fraud, PR gave a voice but no power to opposition movements. But as economic conditions and political unrest began to undermine the position of the Partido Revolucionario Institucional (PRI—the dominant party since 1929), it lost first its absolute dominance of the legislative branch and then, in 2000, the presidency. Around the same time a new formula for public financing of elections came into force, based on votes polled in the previous election. In 1994 the PVEM qualified for 2 percent of the funds disbursed; and 3 percent in 1997.[41] In 1991, the PVEM polled 1.43 percent nationally, and 1.4 percent in 1994, but did not win any plurality seats and fell below the threshold for PR seats. In 1997, however, it polled 3.81 percent and was awarded 8 PR seats.[42] From 1987-2000 its legislators were discounted by the president as among a small handful of leftist opposition parties that had no real effect on legislation. But in 2000, the PVEM joined the Allliance for Change, running candidates for legislative seats in coalition with the *Partido Acción Nacional* (PAN), and supporting the PAN's successful presidential candidate, Vicente Fox. As part of the Alliance the

PVEM increased its seats in the Chamber of Deputies to 15 and in the Senate to 5 from the 1 seat awarded it in 1997. Although data are not available that disaggregates support for the PAN and the PVEM in the presidential contest, the profile of voters for the coalition includes many of the characteristics associated with Green voters in European democracies. In addition to its base among PAN supporters, the Alliance drew support from leftists and moderates discouraged with the PRI and voters that had previously flocked to the more established leftist opposition party, the *Partido de la Revolucionario Democrático* (PRD).

> Fox walked away with the youth vote, especially that of students . . . Fox's advantage over his PRI rival among those with higher levels of education approached three to one . . . Fox also gained the votes of those who paid close attention to the campaign and of those who said the main reason they cast their votes as they did was 'change.'[43]

Mexico is still in the early stages of the transition from a single-party-dominant system. Legislative politics is still an unproven aspect of the developing Mexican democracy, and the PVEM has yet to become a critical partner in any of the major party's governing or electoral strategies. The legislative alliance with the PAN is not an easy ideological match, the PAN being socially conservative and pro-business. But the modification of Mexico's electoral and campaign finance rules have at least temporarily taken the PVEM further than Green parties in most LDCs.

1.3.5. Green Parties and the "New Politics"

In the previous chapter we introduced Inglehart's analysis of changes in values of populations in post-industrial societies. Now we turn to the changes in the distribution of partisan forces which a number of scholars, including Inglehart, believe result from these changes.

The post-World War II agenda emphasized economic growth at any price, and political parties reflected this emphasis with the primary divisions between capitalist and labor parties and class conflict. This fundamental division was challenged by new issues, including environmentalism, and new social groups. By the 1970s, a new left, composed of middle class individuals with post-materialist values increasingly faced a new right, composed of insecure members of the working class. Simultaneously, the public expressed growing skepticism about state planning and control.

It is unclear whether value changes in individuals preceded the rise of new public issues (which is Ingelhart's belief), or if new issues such as environmentalism stimulated the post-modern shift in basic values. What seems clear, however, is that both new values and new issues have reshaped political cleavages within nations of the developed West and led to the formation of new parties.

Inglehart posits a new spectrum to politics cross-nationally, with post-modern values in conflict with fundamentalist values. The rise of Green parties forced changes in the platforms of established parties. Post-materialism inspired a fundamentalist reaction (of those with traditional religious values, among others), a reaction by those who are economically and psychologically marginal and insecure in the face of change. The pattern in Europe and to a lesser extent in North America represented a broad intergenerational cultural shift, a rise of post-materialist issues and a decline of social class voting. Those with higher incomes, education, and occupational status are relatively secure and support the new left-libertarian dimension. The traditional left meanwhile suffered from a diminishing political basis.[44]

The pattern applies less well to LDCs, where the traditional left-right axis may have greater contemporary bearing. The spread of neo-liberalism in the aftermath of the debt crisis of the 1980s has stimulated a reaction in the form of renewed leftist nationalism in parts of the developing world, giving a contemporary face to left-right political cleavages. Recent elections have brought back leftist populism in Venezuela (1998), Brazil (2002), Argentina (2003), Chile (2005) and Bolivia (2005). And although green politics seems not to have been a determining factor in any of these contests, as explained above, ENGOs and environmental movements have made common with the poor, indigenous and disaffected middle classes (especially in Brazil and Bolivia). The exception, Mexico, where the Green party allied with the conservative PAN in 2000 is explained by the widely held perception that the traditional populist (PRI) and leftist (PRD) parties represented the status quo.

The limited electoral appeal of Green parties suggests that the "new politics" of autonomy and democracy has not overshadowed the "old politics" of economic redistribution. It has added another layer to the existing partisan distribution, much as the formation of labor parties in the early twentieth century, and the capitalist reaction to them, added a deep patina to the nineteenth century division between rural conservative and urban liberal political forces.

2. THE MEDIA AND PUBLIC OPINION ON THE ENVIRONMENT

In most countries of the world, public opinion on environmental issues has become an important driver of policy outcomes. In some cases, public sentiment is expressed directly in demonstrations and protests, usually in response to local incidents of pollution and contamination. In most cases, however, public opinion is developed in response to media accounts of environmental problems; then public opinion is measured by pollsters and academic observers and reported back to the public, again through the media. The media thus play a vital role both in the two-way transformation of information and in the development of public agendas on environmental issues.

The public in western nations has been energized by well-publicized environmental crises. In the United States, for example, pictures on TV screens of the Santa Barbara oil spill of 1969 focused attention on environmental pollution as had no prior event and led to the organization and holding of Earth Day 1970, adoption of the National Environmental Protection Act (1970) and establishment of the U.S. Environmental Protection Agency (EPA). The Three Mile Island nuclear reactor accident of 1978 led to development of new regulations for the Nuclear Regulatory Commission, which required emergency planning at commercial nuclear power plants. The 1984 chemical plant disaster which took 5,000 lives of workers and residents in Bhopal, India led to community right-to-know provisions in the Superfund Amendments and Reauthorization Act of 1986. This required industries using dangerous chemicals to indicate types and amounts of chemicals to those living in areas likely to be affected by accidents.[45]

For a generation, polling organizations have queried citizens of post-industrial societies concerning environmental issues. During this period, opinion polls report that substantial majorities in almost all major socio-economic groups support the environmental movement and governmental programs to protect the environment, in all post-industrial countries where opinions have been measured.[46] Percentages supporting "protection of the environment" range from 61 to 71 percent, as compared to rankings from 19 to 32 percent regarding economic development.

In LDCs, media are less well entrenched in state and society, and in authoritarian systems, they are especially subject to censorship. However, reporting of environmental problems is less constrained than news of labor unrest and human rights violations. Available survey research on environmental attitudes indicates greater enthusiasm for economic develpoment and correspondingly lower support for environmental protection. Nevertheless, reporting on environmental disasters and protests have affected political responses in LDCs and authoritarian countries. Guadalupe Rodrigues builds on the sociological concept of issue networks to posit Environmental Protection Issue Networks in which media outlets and reporters specializing in environmental problems may play a key role.[47] The network of local, national and international NGOs that supported the rubber tappers of Acre achieved international publicity for Chico Mendes and his cause. Brazilian authorities and most Brazilians were taken by surprise when his assassination received news and editorial coverage in the *New York Times* and *Washington Post*.[48]

Reaction to the 1986 Chernobyl nuclear disaster was also affected by the media. Coverage from outside of the Soviet Union of increased atmospheric radiation was the first information to reach Soviet citizens of the incident. The state controlled media covered up the accident, offering only this terse announcement two days after it occurred.

> An accident has occurred at Chernobyl nuclear power station. One of the atomic reactors has been damaged. Measures are being taken to eliminate the consequences of the accident. Aid is being given to the victims. A government commission has been set up[49]

The media cover-up is often cited as a catalyst to Glasnost, Perestroika, and the beginning of the end of the Soviet Union.

A recent case from China shows the ability of the regime to control news on environmental protests. On December 6, 2005, some 300 residents of the south China village of Dongzhou, armed with spears, knives, and dynamite, protested a power company's plans to develop a power plant on their land without agreed upon compensation. In addition to forcible seizure of land, the project would install a coal-fired generator, heavily polluting the village. Plans to fill in a local bay as part of the project would ruin a fishery used by villagers for generations; blasting a nearby mountainside for rubble to use in the landfill and filling the bay would threaten biodiversity.[50]

When complaints to authorities and a sit-down protest failed to gain support of county and provincial (Guangdong) authorities, villagers assembled in the town center, confronted by hundreds of police. Without warning, police violently suppressed the demonstration, in the largest use of armed force against civilians since the Tiananmen protests of 1989. The police left 20 protestors dead in automatic weapons fire and at least 40 missing. The regime imposed a blackout on all news about this episode of environmental protest.[51] The New China News Agency reported a skeletal version of the episode only four days later. Then, the regime, which publicizied the arrest of the commander in charge of the police crackdown, announced that "the police were forced to open fire in alarm."[52] Residents reported to the *New York Times* that local officials, in talking with relatives of those killed in the incidents, told them to report that their relatives had been blown up by their own explosives and not killed by gunfire. If they complied, families would receive $50,000 RMB (US$6,193); if not, they would be beaten.[53]

Still the effects of media attention seem to be episodic. In LDCs and authoritarian states, international attention can stimulate local attention (even outrage). However, in both EDCs and LDCs, the opinion survey data leads scholars to question how deep public support is for environmental issues, particularly when they involve regulations to implement programs that increase their personal cost or inconvenience.

3. ENVIRONMENTALISM AND DEMOCRATIZATION

The rise of environmentalism in countries throughout the world has not operated in a vacuum. It has occurred simultaneously with significant economic restructuring through the expansion of global trade, the development of "Third Wave" democratization movements in LDCs, and the onset of post-modernization in EDCs, among other social and political forces. We consider the relationship between environmentalism and democratization first by examining the Third Wave democratization movements. Then we turn to the constraints on environmentalism in authoritarian countries. We conclude the section and chapter by examining the impact of environmentalism on democracy in the EDCs.

3.1. Third Wave Democratization Movements and Environmentalism

The "third wave of democratization" is a phrase developed by Samuel Huntington to describe the 15-year period following the end of the Portuguese dictatorship in 1974 when "democratic regimes replaced authoritarian ones in approximately thirty countries in Europe, Asia, and Latin America."[54] By democracy, Huntington means a political system in which the:

> most powerful collective decision makers are selected through fair, honest, and periodic elections in which candidates freely compete for votes and in which virtually all the adult population is eligible to vote. . . . It (the concept) also implies the existence of those civil and political freedoms to speak, publish, assemble, and organize that are necessary to political debate and the conduct of electoral campaigns.[55]

Our question concerns the relationship between the two processes—whether they are coincidental, causal (and if so, in which direction), or express a correlation dependent on unique aspects of culture, social class structure, historical context, and relationships of power and authority.

It appears that in the Latin American cases of democratization, political liberalization and in turn development of democracy had a causal impact on the development of environmental activism and progress in addressing environmental problems (with the possible exception of Brazil).

Mumme and Korzetz indicate that in most Latin American countries, notwithstanding establishment of environmental agencies and legislation, the authoritarian governments resisted effective implementation. Only at the onset of political liberalization were ENGOs and environmental agencies able to operate freely. Greater citizen awareness of environmental issues and participation in ENGOs as well as other civic organizations led to the strengthening of civil society; in response governments began to implement environmental policies in order to increase their efficacy.[56]

In Eastern Europe and particularly Hungry, Poland and Czechoslovakia, environmental clubs, groups, and activists were major players in the bringing down of communist regimes. They appealed to the younger generation and intellectuals. Growth of these organizations was facilitated by heavy levels of pollution, transboundary pollution, and the 1986 Chernobyl meltdown.[57]

In Southeast and East Asia, the relationship of democratization to environmentalism varied by country. In South Korea, environmental and democratic movements were partners in all three phases of political change: liberalization, democratic transition, and consolidation. Initial protests against environmental pollution also targeted the authoritarian state. Pro-democracy activists often used ENGOs as a shield for their activities. Environmental protesters moderated their confrontational tactics during the democratic transition of 1987-88, and upon democratic consolidation, environmental groups benefited from new members (many pro-democracy activists) and a place on the agenda in electoral campaigns.[58]

In Taiwan, environmental and democracy movement leaders were partners during liberalization, but environmentalists became increasingly skeptical of their erstwhile partners during the democratic transition and consolidation stages. In the Philippines, on the other hand, the struggle against the Marcos dictatorship took center stage, making the environmental movement subordinate to the pro-democracy movement. It became an equal partner with other civil society movements after Marcos' fall from power, and then became important in empowering local communities in carrying out sustainable development strategies.[59]

In the East and Southeast Asian cases, there were differing degrees of democratic consolidation and intensity of electoral competition. However, in the central and eastern European cases of democratization, environmentalism played a more central role in political liberalization and in the democratic transition.

3.2. Environmentalism and Democratization

It is unclear which of these patterns, if any, will be repeated in authoritarian states. To some observers the development of ENGOs in these non-democratic countries appears to be an harbinger of political liberalization. We mentioned previously the rise in the mid-1990s of a number of ENGOs in China. About four dozen ENGOs operate openly in China's major cities, and most have connections to global ENGOs such as Greenpeace, WWF, and Environmental Defense. These organizations are well connected (and sometimes headed by) leading scientists and the Chinese Academy of Sciences; they have linkages to governmental ministries such as the State Environmental Protection Administration (SEPA) and State Forestry Administration (SFA), because they are useful to the state.

Grassroots ENGOs, on the other hand, have traveled a rockier road. Expressing opposition to local cases of environmental degradation, such as large hydropower projects, groundwater/river/lake/wetlands contamination from chemical plants, they have been more likely to engage in direct action, and aroused opposition from the regime. Thus there remain severe constraints on the development of ENGOs, and their use as a conduit for rise of a pro-democracy opposition in countries like China. To the present, ENGOs and the environmental movement have had little direct impact on the embryonic democracy movement in China.

3.3. Expansion of Democracy and Environmentalism in the EDCs

The final question concerns the extent to which ENGOs, Green parties, and environmental movements have "strengthened" democracy in economically developed countries. The question makes an assumption that is not unchallenged: that a high rate of apathy affects the politics of post-industrial societies, and that participation in associations has waned, leading to a decline of civil society as represented in the title of Robert Putnam's best-selling interpretation of modern American society—"Bowling Alone."[60]

It is clear that both ENGOs and the environmental movement represent a growth in civic activity, attracting to public life many individuals who otherwise might have remained withdrawn. However, we lack evidence to determine whether a qualitative change has taken place.

3.4. Sustainable Development and Participatory Democracy

Two final aspects of the relationship between environmental politics and democracy bear mention as they affect both EDCs and LDCs. First, is the role of scientific and communications technologies—especially the Internet and satellite imagery in promoting environmentalism and facilitating the organization of new social movements. In the case of the rubber tappers of Acre, Brazil, international support was stimulated, in part, by the publication of satellite imagery showing the extensive clear-cutting in the Amazon rainforest. The NBA's fight against big dam projects in India has been "globalized" by the use of the Internet for networking among local and international ENGOs.

Second, it is possible that there is a positive relationship between participatory resource management practices and democracy. Scholarship on sustainable development tends to promote the value of stakeholder participation. It is assumed (if not always proven) that the inclusion of affected populations in the formulation and implementation of natural resource management policies lowers the costs and raises the efficacy of resource conservation; and the inclusion of individuals and groups in the policies that directly affect their livelihoods and quality of life is expected to be empowering. Even if democracy goes no further, participatory democracy will have been introduced at levels that are meaningful in the daily lives of participants. This analysis has informed sustainable development policy in LDCs (particularly when participatory practices are made a condition of development assistance). Stakeholder participation is also seen as a way to avoid or mediate disputes among environmentalists and resource users (such as ranchers, loggers and miners) in EDCs. Supporters of this approach see co-management and related techniques as ways around legislated "command and control" regulation that the losers may see as illegitimate. Thus, it is argued that highly localized participatory management introduces democracy to the disenfranchised poor in LDCs, and revitalizes democracy in EDCs where national regulatory processes seem distant, unresponsive and captured by special interests.[61] Research supporting these contentions is still in its early stages. There is a rich, case-study literature describing the formation, design and function of participatory practices at various stages of institutionalization. But we lack substantial cross-national studies of their effects in or on different types of political systems.

---

1 John McCormick, "Environmental Policy in Britain," in Udah Desai, ed., *Environmental Politics and Policy in Industrialized Countries*. Cambridge, MA: The MIT Press, 2002, 121.

2 See discussion of Greenpeace, WWF, and Friends of the Earth in Paul Wapner. *Environmental Activism and World Politics.* Albany, NY: SUNY Press, 1996, 41-151.

3 These are often the proponents of Environmental Justice campaigns discussed in chapter 2.

4 Jonathan Schwartz. "Environmental NGOs in China: Roles and Limits," *Pacific Affairs*, Vol. 77, No. 1 (Spring 2004), 38.

5 For example, in 2004 the Sierra Club membership was 782,000. See Justin A. Dernison, Sierra Club Membership Services, January 24, 2005.

6 See Walter A. Rosenbaum. *Environmental Politics and Policy*, 5th edition. Washington, DC: Congressional Quarterly Inc., 2002, 35.

7 Rosenbaum, 33.

8 Interviews with Fisheries Officers and country representative of the Organization of American States, Roseau Dominica, June 26, 1998 and June 7, 1999.

9 The Goldman Environmental Prize, http://www.goldmanprize.org/prize/prize.html.

10 Japan actively pursues commercial whaling rights through the International Whaling Commission (IWC). SIDS like Dominica receive substantial assistance from the Japanese government, especially in fisheries development. IWC member states each have a single vote

regardless of size. Japan's critics have taken notice." See, Greenpeace International/Oceans Campaign, "Vote-Buying Whistle-Blower Urges Europe to Stop Japan at the Whaling Commission," (26 April 2002), http://archive.greenpeace.org/pressreleases/oceans/2002apr26.html.
[11] Jonathan Rosenberg and Fae L. Korsmo, "Local Participation, International Politics and the Environment: the World Bank and the Grenada Dove." *Journal of Environmental Management*, Vol. 62, No. 3 (2001): 283-300.
[12] "Programme Structure," *the Caribbean Natural Resources Management Institute*, http://www.canari.org/prog.html.
[13] Personal interview with Sandra Guido, environmental interpreter, CIA-Unidad Mazatlan, June 13, 2005.
[14] "History," *Earthjustice*, http://earthjustice.org/about/history.html.
[15] See, for example, Gary Bryner. *Gaia's Wager: Environmental Movements and the Challenge of Sustainability*. Lanham, MD: Rowman & Littlefield, 2001.
[16] See Claus Offe. "Reflections on the Institutional Self-Transformation of Movement Politics," in Russell Dalton and Manfred Kuechler, eds., *Challenging the Political Order: New Social Movements in Western Democracies*. Cambridge: Polity, 1990, 232-50.
[17] Offe, 1990, 239.
[18] Ibid., 248; see also David S. Meyer and Suzanne Stggenborg, "Movements, Countermovements, and the Structure of Political Opportunity," *Americqan Journal of Sociology*, Vol. 101, No. 6 (May 1996): 1628-60.
[19] Bron Raymond Taylor, ed., *Ecological Resistance Movements: The Global Emergence of Radical and Popular Environmentalism*. Albany, NY: SUNY Press, 1995.
[20] Eghosa E. Osaghae, "The Ogoni Uprising: Oil Politics, Minority Agitation and the Future of the Nigerian State," *African Affairs*, Vol. 376, No. 94. (July 1995), 326-7.
[21] Osaghae, 327.
[22] Claude Welch, Jr., "The Ogoni and Self-Determination: Increasing Violence in Nigeria, *The Journal of Modern African Studies*, Vol. 33, No. 4 (December 1995), 635-6.
[23] Greenpeace International, "Ken Saro-wiwa and eight Ogoni people executed: Blood on Shell's hands," http://archive.greenpeace.org/comms/ken/murder.html.
[24] The following relies on Margaret E. Keck, "Social Equity and Environmental Politics in Brazil: Lessons from the Rubber Tappers of Acre," *Comparative Politics*, Vol. 27, No. 4 (July 1995): 409-424.
[25] Paramjit S. Judge, "Response to Dams and Displacement in Two Indian States," *Asian Survey*, Vol. 37, No. 9. (September 1997), 840.
[26] Judge, 840-1.
[27] P.P. Karan, *Geographical Review*, Vol. 84, No. 1. (Jan., 1994), 38.
[28] Interrights Commonwealth Human Rights Law, "Narmada Bachao Andolan v Union of India and Ors [1999] *ICHRL* 141 (15 October 1999), http://www.worldlii.org/int/cases/ICHRL/1999/141.html.
[29] NBA press releases can be found at http://www.narmada.org/nba-press-releases/.
[30] Dates of foundation of Green parties in the other countries are: Finland, Luxembourg, 1979; Germany, 1980; Portugal, Sweden, Ireland 1981; Austria, 1982; the Netherlands, Switzerland, Denmark 1983; Spain, 1984; Italy, 1986; Norway, 1988; and Greece, 1989. See Ferdinand Muller-Rommel, "1: The Lifespan and the Political Performance of Green Parties in Western Europe," in *Environmental Politics*, Vol. 11, No. 1 (spring 2002), 3-4.
[31] See Kitschelt, "The Green Phenomenon in Western Party Systems," in Kamieniecki, ed., *Environmental Politics in the International Arena*, 179-209.
[32] Claus Christian Malzahn, "Happy 25th Birthday Greens. What's the Plan Now?" *Speigel Online*, January 13, 2005, http://service.spiegel.de/cache/international/0,1518,336623,00.html.
[33] Giovanni Sartori, *Comparative Constitutional Engineering*, 2nd edition. New York: New York University Press, 1997, 3.
[34] Ibid., 73.

[35] However, the British Greens sent 2 representatives to the European Parliament after the 1999 elections. Notably, EU elections use party lists and proportional representation.
[36] Ibid., 34.
[37] Ibid., 60.
[38] Muller-Rommel, 7.
[39] Muller-Rommel, 7.
[40] Thomas Poguntke. "7: Green Parties in National Governments: From Protest to Acquiescence?" in *Environmental Politics*, Vol. 11, No. 1 (spring 2002), p. 143.
[41] Mony de Swann, Paola Martorelli, and Juan Molinar Horcasitas, "Public Financing of Political Parties and Electoral Expenditures in Mexico," in Monica Serrano, ed., *Governing Mexico: Political Parties and Elections*. London: Institute of Latin American Studies, University of London, 1998, 156-169.
[42] Alonso Lujambio, "Mexican Parties and Congressional Politics in the 1990s," in Serrano, 170-184.
[43] Joseph L. Klesner, "The End of Mexico's One-Party Regime," *PS: Political Science and Politics*, Vol. 34, No. 1 (March 2001): 107-114.
[44] See Inglehardt, "The Rise of New Issues and New Parties," in Inglehart, ed., *Modernization and Post Modernization*, 237-66.
[45] Rosenbaum, 60-61.
[46] For some American views, see Everett Carl Ladd and Karlyn Bowman, "Public Opinion on the Environment," *Resources* 124 (summer 1996), 5; and The Gallup Poll, "Environment," *The Gallup Poll Organization* (Princeton, December 25, 2000); available at www.gallup.org/
[47] Maria Guadalupe Rodrigues, "Environmental Protection Issue Networks in Amazonia," *Latin American Research Review*, Vol. 35, No. 3 (2000): 125-153.
[48] Keck, 415-20.
[49] Moscow Radio on 28 April, 1986, quoted in BBC News on-line, "Media Recalls Chenobyl," (April 26, 2001), http://news.bbc.co.uk/1/hi/world/monitoring/media_reports/1298257.stm.
[50] Howard W. French, "Protesters Say Police in China Killed up to 20," *New York Times*, December 10, 2005.
[51] French, "Beijing Casts Net of Silence Over Protest," *New York Times*, December 14, 2005.
[52] Joseph Kahn, "Military Officer Tied to Killings is Held by China," *New York Times*, December 12, 2005.
[53] French, "China Pressing to Keep Village Silent on Clash," *New York Times*, December 17, 2005.
[54] Samuel P. Huntington, *The Third Wave: Democratization in the Late Twentieth Century.* Norman, OK: University of Oklahoma Press, 1993, 21.
[55] Huntington, 7.
[56] Stephen P. Mumme and Edward Korzetz, "Democratization, Politics, and Environmental Reform in Latin America," in Gordon MacDonald, Daniel Nielson and Marc Stern, eds., *Latin American Environmental Policy in International Perspective.* Boulder: Westview Press, 1997.
[57] See Susan Baker and Peter Jehlicke, "Dilemmas of Reform," *Environmental Politics*, Vol. 7, No. 1 (1998), and Lillian Botsheva, "Focus and Effectiveness of Environmental Activism in Eastern Europe," *Journal of Environment and Development* (September 1996), 295-96.
[58] Soo Hoon Lee, Hsin-Huang Michael Hsiao, Hwa-Jen Liu, On-Kwok Lai, Francisco A. Magno, and Alvin Y. So, "The Impact of Democratization on Environmental Movements," in Yok-shiu F. Lee and Alvin Y. So, editors, *Asia's Environmental Movements*. Armonk, NY: M.E. Sharpe, 1999, 230-51.
[59] Lee et al., 1999, 241.
[60] Robert D. Putnam, *Bowling Alone* (New York: Simon & Schuster, 2000).
[61] Robert F. Durant, Daniel Fiorino, and Rosemary O'Leary, eds., *Environmental Governance Reconsidered: Challenges, Choices, and Opportunities.* Cambridge, MA and London: the MIT Press, 2004.

CHAPTER 4. POLITICAL INSTITUTIONS AND THE ENVIRONMENT

This chapter examines the structure and organization of the state itself, and the role that its political institutions and arrangement of authorities and powers play in the development and implementation of environmental policy. The relationship between government structure and policies is close. A policy such as reduction in carbon dioxide emissions becomes effectively public only when adopted, implemented, and enforced through government institutions. Institutions give needed specificity to environmental policy. They establish procedures for the making and remaking of policy, and assign responsibility for implementation. To the extent the state is legitimate, its institutions give legitimacy to environmental policy. To the extent it is not, institutions provide coercive powers to enforce policy.

After discussing briefly the nature of institutions, we consider four dimensions of institutional authority cross-nationally: constitutional versus authoritarian systems, the territorial distribution of authority, concentration of decision-making authority, and differences in judicial institutions. The chapter concludes with broader observations on the policy-making and political opportunity structure of nations.

1. THE NATURE OF INSTITUTIONS

Political institutions can be defined broadly as formal rules, compliance procedures, and standard operating procedures to shape strategies, goals, and actions of social actors. When discussing institutions, what first comes to mind are the executives, legislatures, courts, and bureaucracies of the modern state. These all are embodiments of important and legitimate purposes of the state such as the making of law; typically in EDCs they have sufficient power to accomplish the purposes of environmental protection; and they persist over time, outlasting human lives.

We have already used institution to refer to the electoral systems of modern states. These are more in the nature of formal rules of procedure for counting votes and assigning seats in legislative bodies and executives. Like the more familiar organs of government, however, they too influence the behavior of people and shape their goals and objectives.

Institutions are also repositories and sources of political power. We have already discussed the ways that organized actors attempt to influence or even create institutions in pursuit of their own interests. An environmental movement, for example, may stimulate the creation of new state institutions such as the environmental protection agencies that proliferated in the 1970s. These institutions took a central role in the implementation of environmental policies and became sources of scientific and administrative expertise in EDCs. Therefore, control of these institutions is a goal in the struggles over environmental policy. Movements may also lead to the formation of Green parties which may influence policy through electoral and legislative strategies. As in the German case, these organizations too become contested.

The institutional focus of comparative political analysis has shifted over time. Behavioralists considered different institutional arrangements as products of different political cultures and levels of development. A more recent line of inquiry called "neo-institutionalism" suggests that political institutions themselves shape strategies and goals of social actors in ways that can dramatically alter outcomes. The new institutionalism also comes in a rational-choice variant where institutions are seen as important in solving collective-action problems, agency problems, reducing transaction costs when the number of participants is high, and limiting principal-agent problems by overseeing and enforcing contracts, including those of environmental policy.

Collective action problems are prevalent in the area of environmental politics, at both the domestic and global levels. They refer to situations where the activity of the state or an international regime (or institution) is required, in order to resolve problems extending beyond the capacity and reach of single individuals, groups, or regions. Examples include addressing factors causing air pollution or toxic contamination in aquatic ecosystems. Because most of the effects accrue downwind or downstream, those who cause or contribute to the pollution have no incentives to remedy it. They are most likely to apply short-term cost-benefit analysis to the problem and conclude that it is not in their interest to absorb the costs of "cleaning up their act" when someone else is "paying the price." Furthermore, cleaner air and water are indivisible benefits (if they exist for one they exist for everyone). Rational polluters will free-ride on the anti-pollution efforts of downwind and downstream communities (which will certainly be suboptimal), until an institution with a comprehensive jurisdiction is created. In other words, the state must imbue an institution with the authority to assess the costs of environmental protection and make those assessments binding through its powers to tax, fine and license.

Principal-agent problems also are prevalent in environmental politics. They refer to situations concerning representatives (agents) of interests (principals), who may be individuals collectively, business firms or other organizations. The agents may act on bases other than the interests to be represented, including their own preferences and goals. For example, one principal-agent problem in environmental politics is when the agent of constituents, such as members of Congress in the United States, delegate authority to administrative agencies when they can as easily craft specific instructions. This permits the agencies to form alliances with the business firms being regulated, with results antithetical or at least different from legislative intent.[1]

Transaction costs refer to the expenses which individuals, business firms, or organizations pay in order to meet requirements or do business with others. Business firms often seek out environmental regulations because a stable regulatory regime reduces uncertainty about the costs of pollution and may, over time, reduce government intervention (and threats to property rights). Indeed, one stream of theoretical interpretation suggests that environmental regulation reduces transactions costs over time. This may seem counterintuitive since environmental regulations can increase the cost of doing business. Businesspeople frequently complain about the expenses of compliance: the time, the paperwork, the training, the new technologies and more expensive materials and fuels required. But these complaints mainly reflect short-term expenses to established and/or smaller businesses confronting new regulations.

Internationally, a great deal has been written, mostly anecdotal, about environmental regulations driving businesses to "pollution havens"—regions and states that attract businesses by offering lax environmental regulation. The argument assumes that LDCs will engage in a "race to the bottom," lowering environmental standards in a competition to attract foreign investment. Critics of corporate policy frequently cite the relocation of assembly plants (*maquiladoras*) from the U.S. to northern Mexico to illustrate the very real environmental effects of the southward movement of manufacturing.[2] But the centrality of environmental regulation in the relocation of industry is open to doubt.

> Numerous studies have concluded that, in comparison with other factors considered by businesses, pollution control costs are not major determinants of location. More important variables include distance to market, infrastructure quality and cost, and so on. In a study of Mexican maquiladora plants, Grossman and Krueger have found that pollution abatement costs were not a major determinant of imports from Mexico, while the cost of unskilled labor was of paramount importance.[3]

Large, multi-national corporations based in EDCs also benefit from standardization and economies of scale. It is usually more cost effective to export cleaner technologies developed to comply with home country regulatory standards than to maintain a separate set of dirtier technologies in overseas plants.[4] Thus far, research seems to indicate that factors other than environmental regulations—especially labor costs and proximity to markets and supply chains—drive the foreign direct investment decisions of MNCs. If there is a "race to the bottom," therefore, it has more to do with the conditions for working people than the environment *per se*.

2. CONSTITUTIONAL LIMITATIONS ON STATE POWER

The range of governments globally covers those in which no limits are imposed on leaders, singly or collectively, to those that are highly constrained by constitutions. A constitution refers to the basic rules about decision-making, rights, and the distribution of authority in a state. It may be a specific document setting out principles, such as the U.S. Constitution written in 1787. More generally, a constitution is a basic set of rules and principles, whether written and available in one document or a set of customs, practices, and laws, such as the "unwritten" British Constitution. Constitutions are particularly important in political systems based on the *rule of law*, which applies generally more to post-industrial EDCs than to the economically developing countries.

Constitutions define the sources of authority and range of state power. There is no absolute requirement that a constitutional system be a democracy. One can imagine a system in which a designated group of political elites wield uncontested power according to clearly codified rules that circumscribe the realm of their authority. But democratic theory tends to conflate constitutionalism and democracy because constitutional limitations on state authority leave important areas of social interaction and decision-making to free individuals and groups (i.e., civil society). Thus, constitutional systems tend to spawn more vibrant environmental movements. But how influential movements will be, and how effectively they will overcome the problems outlined above—collective action, principal-agent, and transaction costs—depend on the institutional characteristics that we now consider.

2.1. Liberal State Systems

Constitutional systems provide limitations on state power and authority, with protection of civil rights against government interference except under specified circumstances. Rights to assemble, petition, speak, and publish are available to all individuals, groups, and interests which seek to change or defend policy. They are "liberal" because they emphasize that individual consent is the foundation of state power and individual rights must be secured by governments. Such rights are essential to the development of environmental interest groups, movements, and political parties. Without them, the environmental group or party is subject to change at the whim of government officials and administrators. Liberal constitutional systems also provide multiple ways for citizens to hold officials accountable for their conduct in office: regular elections; constitutional methods for removal from office (impeachment and votes of confidence and censure); and laws that apply to state agencies and officeholders as well as business entities and individuals.

Constitutions also may embed environmental protections in the constitution itself. In liberal state systems, constitutional protection of the environment gives individuals, groups, and movements an easily enforceable right. In the absence of constitutional rights to environmental protection in liberal state systems, other constitutional protections, such as those to equal protection of the laws or due process, must be employed.

2.2. Non-liberal State Systems

Policy-makers in authoritarian systems are selected by military councils, families, oligarchies, and dominant political parties. They may have originally come to power via revolution, rebellion or coup d'etat over-throwing failed democracies or debilitated and corrupt autocracies. Many modern authoritarian regimes have come to power ostensibly to carry out a development strategy. This was true not only of the Marxist revolutionaries of the Soviet Union, China, and Cuba, but of the military dictatorships that swept Latin America, Southeast Asia and Africa in the 1960s and 1970s. Leaders of these regimes see their reign as revolutionary or transformational; dedicated to overcoming the consequences of imperialism, exploitation, corruption, and elitism. Of necessity, their states engulf much of society, at least temporarily, for the overall purpose of "catching up" to and surpassing the exploiters and/or to discrediting the old regime.

Most non-liberal states also have constitutions, but they are usually only marginally relevant to the actual distribution of political power, and can be changed with each power shift in the regime. For example, since its establishment in 1949, the People's Republic of China has adopted four constitutions, in 1954, 1975, 1978, and 1982 (the most recent, embodying reforms of Deng Xiaoping). Amendments to the 1982 constitution have enshrined environmental protection among the basic rights of the people, and this has signaled to institutions, such as the state court system, that hearing complaints against environmental degradation is acceptable.[5] The issue in the Chinese state context is whether the constitutional provision will protect against economic development objectives of the state which conflict with environmental protection.

The Mexican Constitution of 1917 codified the political, economic and social aspirations of the surviving factions of the Revolution of 1910. It contains provisions for an agrarian reform that respects traditional peasant cooperatives (*ejidos)* and requires natural resources to be managed in the national interest. Historically, these provisions have been selectively enforced and only to the extent they served the interests of presidential power and the dominant party.

Cuba promulgated its first post-revolutionary constitution in 1976 (17 years after the Revolution), establishing new formal mechanisms of representation but effecting no real change in the distribution of political power, leadership, or the dominance of the state over society. Grassroots and local environmental initiatives are not unheard of in Cuba, and there is variation in approach and effectiveness at the provincial and municipal levels. But the ability to effect change still depends on the consent and support of the national government which alone can determine national development strategies.[6]

2.3. Constitutions and Environmental Law

Whether the constitution mentions the environment or the concept of sustainable development may facilitate the development of policies protecting environmental values. Of greater importance are provisions found in constitutions establishing a national and municipal system of law. The constitution establishes the jurisdiction over which national law operates, and, as we shall see below, whether sub-national units (provinces or states and municipalities) have the legal capacity to co-manage state functions such as environmental protection.

A traditional question of comparative law, resolved in constitutions, is whether the government follows a civil, common, or socialist law tradition.[7] This influences the role of courts in the state's political system, as well as the uniformity of law across different jurisdictions. The constitution also influences the relationship which will obtain between substantive and procedural environmental law, and the ease with which national environmental laws can be integrated and harmonized with international environmental norms. Constitutions also help define property rights, establishing varying levels of protection for private ownership. Even the most liberal of constitutions will include a right of eminent domain, and state authority to regulate the use of private property especially as it affects public interests and third parties; while socialist constitutions typically emphasize public benefit and collective ownership.

3. TERRITORIAL DISTRIBUTION OF AUTHORITY

A basic decision rule established in constitutions is the geographic distribution of authority between the central (national) government and sub-national levels, such as states or provinces, and municipalities.[8] At one extreme are unitary systems, like Japan, France, China, and Egypt, the most highly centralized, with power and authority concentrated in the central government. At the other extreme are confederal systems, which assign ultimate power to states or provinces. The European Union (EU), although a super-state, is confederal in structure, as defense and foreign policy powers remain lodged within component nation-states. In the middle are *federal* systems, like the U.S., Canada, Russia, India, Nigeria, and Brazil, in which both central and state/provincial governments have separate spheres of authority and the means to implement their power.

It is important to keep in mind the distinction between formal constitutional arrangements and actual distribution of power in states. In some unitary systems, regional and local units may acquire power that the central government rarely challenges; in some federal systems, such as the United States, power may steadily migrate from the states to the center. For our purposes, the issue is the amount of centralization of authority, which presumably influences the integration of environmental policy nationally and the consistency in implementation.

3.1. Federal and Confederal Systems

Only 25 nation-states in the early twenty-first century have federal systems, a minority of the world's governments. Two factors tend to explain the evolution of federal systems: the attempt to resolve problems of power and administration in territorially vast nations, and the attempt to dilute power of ethnic, linguistic, racial, and cultural minorities. Three of the world's four territorially largest states are federal systems—Russia, Canada, and the United States, and about 40 percent of the world's land areas are found within federal states. Most of the other federal states—for example, India, Nigeria, Australia, Brazil, and Germany—cover large land areas too.

Federalism is also a form of government used to reduce the potentially divisive impact of minority groupings. The argument is expressed well in James Madison's *Federalist #10*, in which he argues that the great threat to free peoples is the domination of politics by factions, which express the passion of a minority and not the general interest of all. Although Madison used "faction" to discuss economic groupings or classes, it can be applied equally well to religious, cultural, linguistic, ethnic, and racial groups. A federal system of government, said Madison, would spread factions out across a large terrain and encapsulate them within geographic regions, making it difficult for them to unite nationally.[9] Nigeria has undergone several constitutional reforms since gaining independence in 1960. Each time the number of states was increased in hope of ameliorating sectarian violence.

Federalism also is a governmental system that in theory allows sub-national units to address problems not general to the whole body politic, and to experiment in their resolution. The apparatus of the central government is duplicated in each of the sub-national states or provinces, and each may have sovereign powers. Of course variation across federal systems is great with respect to the amount of centralization. In the United States, for example, states have "residual" powers, not spelled out in the federal Constitution. Congress, on the other hand, beyond the inherent powers of government has specified powers, such as to regulate interstate and international trade, which have been progressively liberalized through Supreme Court decisions such as *McCulloch v. Maryland (1819),* which effectively added an "elastic clause" to the Constitution. In contrast, the Canadian federal system has only ten provinces (as compared to 50 U.S. states and 87 Russian republics), and each owns and manages Crown lands and their resources.

Federal systems also differ with respect to the amount of equalization of revenue among the states or provinces, which directly influences implementation of environmental laws and policies. In both the U.S. and Canada, the federal governments have greater tax capacity than state or provincial governments. Yet the Canadian government applies an equalization grant scheme to reduce disparities of provincial government income. As noted in chapter 3, Nigerian states have come into conflict with federal authorities over the distribution of oil revenues.

Federal systems differ as well in the amount of decentralization incorporated in the basic division of central/sub-national powers and responsibilities and in the extent to which the central government decentralizes programs directly to municipal or local governments. For example, in the United States, federal air and water pollution legislation is shared between federal and state governments. The national Environmental Protection Administration, with its headquarters in Washington, DC and its ten regional offices works cooperatively with states, most of which issue the relevant permits. Nearly 40 of the U.S. states have the authority to issue water permits and all but one state issue certified pesticide permits.[10]

Between federal and confederal structures is an additional type of governing arrangement called consociationalism. Consociational systems result from national power sharing arrangements designed to settle ethnic, religious, linguistic, territorial, and/or racial conflicts among communities. Switzerland with its rotating collegial national executive and four distinct linguistic regions is the most stable example of a modern consociational state. Policy-making is highly decentralized, with the canton (main sub-national unit) being the center of most political decision-making.[11] Other examples include Belgium and the notably less successful case of Lebanon.

The world's best example of a confederal system is the European Union. Each of the 25 member states either sits on or is represented in the primary institutions composing the EU:

- The Council of the European Union (whose members are heads of government, meeting twice annually), but usually the executive focus is on
- The Council of Ministers (which includes a Council of Environmental Ministers, meeting at least twice annually);
- The Eurpean Parliament (representing states proportionally; it has an environmental committee);
- The Commission (the EU's bureaucracy with a Directorate General responsible for environmental legislation and with supervisory responsibility over the new European Environmental Agency); and
- The European Court of Justice.[12]

Initially, each state had veto power over legislation affecting it, the defining characteristic of confederal systems; and when, under the Single European Act of 1987, the resolution of environmental matters was adopted into the Treaty of Rome, the requirement for unanimous agreement remained. However, an exception allowed use of qualified majority voting (which gives the larger states more votes than the smaller ones) in the Council of Ministers on environmental matters related to completion of the single market. The process, called the "cooperative procedure" further empowered the European Parliament. The Treaty on European Union (Maastrich Treaty) taking effect in 1993 extended qualified majority voting to most environmental matters and introduced a co-decision procedure.[13] The Treaty of Amsterdam of 1999 makes the co-decision procedure the norm. Thus, the confederal arrangement of the EU has been transformed increasingly into a situation resembling a centralizing federal system, such as Germany.

3.2. Unitary Systems

Most of the world's nation-states are unitary state systems, in which the national government is the sole source of sovereignty. Virtually all of the unitary systems, however, have provinces, prefectures, or other sub-national levels of government, and they have municipal levels of government as well. However, the central government may transfer to or withdraw powers from these sub-national governments at any time.

In some unitary states, national legislation has defined significant "home rule" powers for provinces and municipalities. For example, in 1979 Denmark transformed its colony Greenland into a home-ruled jurisdiction, effectively making it a province with autonomous powers for local legislation (Denmark retains foreign relations and economic development authority). In the area of environmental policy and implementation, there are few differences between the practice of Greenlandic authorities and those of Canadian provincial authorities. In contrast, decentralization of political authority has been impractical in the Japanese context. Prefectures, administrative units created to replace the feudal daimyo after the Meiji Restoration, coordinate the activities of cities, towns, and villages. The Japanese state adopts environmental legislation, and its implementation is a shared responsibility between the central and local governments. While the central government defines environmental policy and sets standards, prefectural governments are responsible for monitoring.[14]

A number of unitary states have practiced *devolution* of authority to provinces or other sub-national levels. The central government transfers the powers in specified areas to provincial or local authorities and may, as in the British experiments with devolution of authority to parliaments in Northern Ireland, Wales and Scotland, allow substantial autonomy. As of 2000, the Welsh and Scottish parliaments have responsibility for economic development, tourism, and environmental policy. The Chinese government has the appearance of considerable devolution in the number of autonomous regions (such as Tibet and Inner Mongolia), autonomous zhou (districts), counties, townships, and villages, but until the onset of economic reforms they had little autonomous decision-making powers.

The Dengist reforms in China led to considerable devolution of economic decision-making to provinces and local governments, based on the premise that greater flexibility was needed to spur economic development. This extends to the area of environmental protection, where most of the implementation (and resources required to put policies into effect) occurs at provincial and lower levels of the government bureaucracy. Lieberthal notes: "Upper levels of the bureaucratic system are constrained in their ability to intervene to force local leaders to take account of the larger environmental costs of their actions."[15] When there are strong pressures to develop an industry that may impinge on environmental preservation values, the contradiction between central and local control becomes most clear. Many of the recent stories of habitat degradation in China feature a provincial or local government, seeking to develop the local economy, in conflict with a national law and environmental office.[16]

3.3 Does Federalism Make a Difference?

This review suggests that administrative decentralization and/or devolution may bring about the same effects as federalism. A few comparative studies have tested the effect of federalism on environmental performance of nation-states. For example, Lundqvist asks whether differences in political structures, including federalism, cause differences in selection of environmental policy alternatives. He designs a comparative study of three countries—the U.S., Canada, and Sweden—which are "most similar" with respect to the dependent variable (controlling air pollution) but different in political structure.

Lundqvist finds that Canada's Clean Air Act of 1971 uses an Air Resource Management (ARM) approach which establishes criteria for quality of ambient air. Sweden's Environmental Protection Act uses the Best Available Means (BAM) approach, which establishes standards for maximum permissible concentration of air pollutants at the point source. The U.S. originally followed the ARM approach, but with Clean Air Act Amendments, combined this approach with BAM. The Canadian policy authority is most decentralized, while the U.S. increasingly gave power to the federal government. Policy powers are most centralized in Sweden.[17] This appears to follow the degree of decentralization within geographic distributions of power: Canadian provinces have strong decisional authority as compared to the weaker U.S. states, and Sweden is a unitary state system. However, other factors, such as degree of judicial involvement and political culture, also influence the relationships.

A second perspective on the issue of federalism's relevance to environmental protection comes from a study of Kenyan approaches to sustainable development. Orie asks what potential effects federalism would have on environmental management in the context of developing multipartyism. He compares the likely impacts of cooperative/coordinate federalism (where component governments would have some degree of autonomy) with organic federalism, where the central government would play the dominant role. He argues that organic forms of federalism (closer to unitary than federal rules) would best conduce to reflective management of the environment in nations like Kenya, where not only are the institutional and jurisdictional framework organizationally fragmented, but also there is fragmentation along ethnic, tribal, and regional lines.[18]

A recent study of environmental policy and performance in industrialized countries, draws hypotheses on the role that political institutions, including federalism, play. Desai analyzes the treatment of seven states; four are federal (the U.S., Canada, Australia, and Germany) and three unitary (Britain, Italy, and Japan). Based on descriptions and analyses of chapter authors, he hypothesizes:

> [F]ederal systems with a long tradition of a weak central government, winner-take-all electoral systems, developmentalist ideology, and states or provinces economically heavily dependent on exploitation of natural resources are likely to be characterized more by conflict than by collaboration among national and state or provincial governments. On the other hand, federal systems with a neocorporatist policymaking system and a proportional representation electoral system are likely to have more collaborative federal-state relations. In a nonfederal unitary system, the local government is likely to represent its citizens' voice for local environmental protection. Local governments are likely to be an important influence on the central government for stronger policy actions to protect the environment.[19]

In short, federalism alone is not associated with large differences in environmental policy outcomes. However, in combination with other factors, such as electoral institutions, political culture, and corporatist structure, it does make a difference.

4. CONCENTRATION OF DECISION-MAKING AUTHORITY

A second basic decision rule is the *separation of powers* among different branches of government. Authoritarian states do not permit the possibility of separation of powers. Legislatures and courts may serve as sounding boards for opposition movements, occasionally with great political effect, but the influence on formal government decision-making is nil. For democracies in the post-World War II era, two forms of government have drawn most attention from comparativists: presidential and parliamentary systems. The democratic presidential system provides two separate agencies of government—the executive and legislative—separately elected and authorized by the people. The two branches have fixed terms and specified powers; they cannot easily unseat one another. The parliamentary system, on the other hand, makes the executive and legislative branches interdependent, with the cabinet emerging from the elected legislature and chaired by a prime minister (or premier) who heads the government and selects other cabinet members.

However, the world of states is not neat, and there also are hybrid types often called "semi-presidential," where the president has powers to dissolve the legislature and call for new elections. These types of arrangements have become increasingly important. The longest-lived example is the French 5th Republic, created in 1958 to stabilize a parliamentary system by adding a strong, elected executive and a majority electoral system. It has been emulated in the Russian Constitution of 1993, the constitutions of newly democratizing Eastern European states and in Taiwan. In France, where multipartyism and strong parliaments are engrained in political culture and tradition, the distribution of power between the president and the premier and cabinet has wavered. In Russia, presidential supremacy has been the rule.

The significant environmental policy issue is the amount of *concentration of powers* in the state, ranging from authoritarian to parliamentary through hybrid forms to presidential systems. We examine some likely impacts of type of concentration, presidential versus parliamentary, and then consider a special form of concentration, corporatism.

4.1 Presidential Systems

The separate election of a president and his/her ability to focus national attention on environmental issues through campaigns and opinion leadership, and influence on the legislature and broader public give presidents opportunities of national leadership in environmental policy unrivaled by prime ministers in parliamentary systems. Also, presidents may have greater opportunities to influence bureaucratic action than do prime ministers, because of their ability to mobilize public opinion. However, this assumes that presidents are of the same party as the legislative majority, which may not be the case as seen in American governments of the 1980s and 1990s.

Presidents can also be hampered by their need for legislative ratification of treaties. After signing the Kyoto Protocol on greenhouse gas reduction in 1997, President Clinton faced such overwhelming bi-partisan opposition that he did not even submit it to the Senate for ratification. The low party discipline associated with presidential systems, the effect of interest group lobbying, and a constitutional requirement for a two-thirds Senate majority to ratify treaties make U.S. participation in international environmental accords difficult.[20]

It is the greater opportunities for conflict between executive and legislature which lead to the generalization of greater ease in development and implementation of environmental protection in highly concentrated than in dispersed authority systems. There are important exceptions to this, however. A popular president can use his/her appeal with the electorate to sway a recalcitrant legislature, even one controlled by a different party, as was the case with Ronald Reagan. And presidents affect policy implementation through their power to appointment cabinet secretaries/ministers, and agency heads, and their influence on civil servants.

4.2. Parliamentary Systems

Parliamentary systems are fused types of governmental authority, with prime ministers or premiers arising out of the majority or coalition having won most seats in the national legislative body. The leader's position of authority is predicated on his/her ability to hold the coalition together, and stability of government may be a somewhat greater problem in parliamentary than in presidential systems. Once the government is formed and embodies a stable interest, however, it is able to achieve legislation on its policy agenda.

The party and electoral systems play a critical role in the stability and efficiency of parliamentary governments. As we have seen in the case of the German Red/Green coalition, governing alliances are stabilized not only by ideological affinities, but by awarding important cabinet ministries to coalition partners. Typically, party discipline is stronger in parliamentary systems since failure of a key piece of legislation can bring down a government. When a parliamentary system is combined with a majority electoral system, as in the British case, coalition governments are uncommon and governments can be even more stable than in presidential systems.

Little comparative research has been done on the impact of presidential versus parliamentary systems on environmental policy outcomes. One such study is David Vogel's 1993 article which treats differences between separation of power and parliamentary systems as related to type of government.[21] Rothenberg summarizes these findings:

> [P]arliamentary leaders are more likely to be blamed for failing to keep commitments for collective goods such as the environment than those in separation-of-power systems and are better at representing diffuse interests and resisting concentrated ones such as polluting industries while making policy. However, if they do not represent concentrated interests, such interests are not likely to have much political effectiveness. By contrast, presidential systems, given the separation of powers, provide diffuse interests with many places to influence policy. Because there are multiple points at which diffuse interests can try to affect policy, there is a greater likelihood than in parliamentary systems that interest groups ignored by the executive branch can, nevertheless, be influential.[22]

In general, this suggests that ENGOs will be happier in separation of powers than parliamentary systems.

The research literature gives greater emphasis to "consensual politics," which is unclear as to its institutional causes, and fragmentation and decentralization, which are clear conceptual references to federalism and devolution experiments. Desai does note that "Where the center is strong, environmental policy is likely to have a clearer direction, though not necessarily a pro-environmentalist direction."[23]

4.3. Corporatist Systems

In chapter 2 we identified corporatism as a particular alignment of the interests of business and labor, whose representatives sat at the table with agents of the state in the determination of national economic policy. But concepts of corporatism and pluralism are institutional factors too. They specify the *type of interest representation* in the nation-state, which may affect levels of environmental degradation.

As mentioned, corporatist institutions describe a system of interest representation in which a small number of strategic actors (invariably capital, sometimes labor), which are organized in peak associations, represent most of the population in an "encompassing" fashion. In authoritarian corporatism, membership is usually mandatory since it provides a hierarchical structure of control, allows the state to more efficiently coordinate development strategies, and economizes on the use of force to maintain order. In democratic corporatism peak associations enjoy substantive input in the decisions that affect their members.

Only in a few democratic corporatist countries, such as Norway, has the environmental interest been expressed in this institution of strategic policy-making. Pluralist institutions, on the other hand, describe a large number of atomistic interest groups engaged in a competitive struggle to influence national policy. Corporatism is relevant to the issue of concentration of powers, in that the acknowledged corporatist systems (for example France and Germany) are hybrid or parliamentary and not presidential systems.

In authoritarian and transitional corporatist systems environmentalists have difficulty overcoming the entrenched, clientilistic relationships between peak organizations and powerful party and state officials that dominate policy-making. Patron-client networks that formed to implement strategies of rapid development and channel political support for regimes do not yield easily to oppositional groups with competing values. Clientilistic networks, such as those built by the PRI and its constituent organizations in Mexico, tend to persist through the transition to democracy. At state and local levels they continue to control the distribution of political goods even as changes occur in national politics. Therefore, environmentalists will have difficulty penetrating the institutions of policy-making. The two options we have previously examined for Mexican environmentalists, direct action and national-level party politics, have achieved little to-date. Police, the military and party bosses resist direct action, as in the case of the Organization of Peasant Ecologists in Guerrero (see chapter 2). And protestors get little redress from courts, which are formally independent but limited by institutional weakness and corruption. Grassroots organizations succeed only when conditions are ripe, such as during the economic crises of the 1980s and 1990s that delegitimized peak organizations and party leaders, and slowed the rate of economic development. But even in such instances grassroots environmentalists have had to make extraordinary efforts to develop and maintain alternative policy networks.[24]

5. COURTS AND A "RIGHT" OF THE ENVIRONMENT

The final institutional factor we consider is the judiciary and the degree of its independence. In some nations such as Great Britain, China, Egypt, and Mexico, courts lack the power to limit coercive government action against citizens. In Great Britain, for example, there are few court challenges on environmental issues, because formally the courts are not separate from the parliament.

In other nations, such as the United States, Japan, Germany, and India, courts may be independent and able to protect the rights of citizens and also police other parts of the government to see that their powers are properly exercised (a power identified as *judicial review*). Independent court systems may be powerful agents in the implementation of environmental protection legislation against business interests and reluctant bureaucracies, but as is the case with other political institutions, this varies by partisan and interest group factors. And in some LDCs the effectiveness of courts is limited by corruption and intimidation.

In the United States, the 1970s were an era of advantageous federal court decisions in both substantive and procedural issues on the environment. As Rosenbaum notes:

> The federal courts greatly expanded opportunities for environmental groups to bring issues before the bench by a broadened definition of "standing to sue," a legal status that authorized individuals or organizations to sue governmental agencies for failure to enforce environmental legislation.[25]

Indeed, some environmental legislation in the United States, particularly the Endangered Species Act (ESA), gives legal standing to any person who "may commence a civil suit on his own behalf . . . to enjoin any person, including the United States . . . who is alleged to be in violation" of the ESA.[26]

However, as business interests formed their own public-interest law legal foundations, and as the Reagan and Bush (both I and II) administrations appointed conservative judges to the federal courts, environmental litigants did less well. As legal specialist Wenner noted about the Reagan administration: "Reagan policy makers reduced the severity of many of the regulations originally drafted by regulatory agencies. Many of these new regulations have now been appealed to the courts, and there have been few victories for environmental groups."[27]

Too, the independence of courts is subject to influence of strong executives using extraordinary means. During the 1970s in India, the Supreme Court conflicted with Indira Gandhi's government, especially when it sought to limit the court's power to protect property rights. Under a 1975 declaration of emergency rule, the Indian Parliament amended the constitution to prohibit the court from reviewing amendments to the constitution. Ginsburg notes: "In the face of these attacks on jurisdiction and threats to judicial independence, the court largely submitted to politicians' desires."[28]

Countries with independent courts thus have an additional venue in which environmental complaints can be processed and perhaps addressed. As political institutions, however, the ideological composition of courts and the nature of the pressures they encounter will vary over time. An autonomous judiciary, such as that in Germany, may not provide a protected arena for environmental litigation. Dryzek et al. comment:

> While environmental groups in Germany can seek redress through the legal system, German activists have had to get by without some of the legal instruments long taken for granted by their US counterparts. For example, there has been no strongly established right to freedom of information, and class action suits have not been part of the federal legal system.[29]

The European Court of Justice (ECJ) is an autonomous institution of the EU. Through direct actions such as appeals and referrals from domestic courts, the ECJ clarifies and interprets EU law.[30] Its rulings take priority over domestic law and are directly applicable to citizens of member states. Too it arbitrates differences in areas of responsibility and competence of EU institutions. It thus ranks along with the more independent courts of nation-states internationally.

6. ENVIRONMENTAL POLICY-MAKING

Although nations vary greatly in their capability to effect environmental outcomes, they all have a system to generate national policies, including those on the environment. Here we introduce the nature of the environment as a policy problem, the types of networks engaged in policy-making, and the stages of the policy-making process.

6.1. Nature of the Environment as a Policy Problem

Environmental degradation such as air pollution is one kind of problem confronting the state. It differs from other problem areas, for example foreign affairs, health, labor, education, or energy in predictable ways. Those differences may be critical elements in the process of policy-making.

Scholars differ in description of the core components of the environment as a policy area or issue. Carter compiles what is perhaps the most exhaustive list including seven characteristics: public goods, transboundary problems, complexity and uncertainty, irreversibility, temporal and spatial variability, administrative fragmentation, and regulatory intervention.[31] In our view, only three of these characteristics clearly differentiate the environment from other policy areas, specifically public goods, uncertainty, and irreversibility. Other policy areas and particularly foreign policy reflect transboundary problems. Each policy area has elements of complexity, which certainly is the case of social policy. Variability across time and space is an important dimension of energy policy. Administrative fragmentation applies to all of the relatively recent problems encountering modern states, and is no truer of environmental policy than population policy or consumer protection. Finally, regulatory intervention is as true of labor, transportation, and health policy as it is of environmental policy.

The remaining three characteristics—public goods, uncertainty, irreversibility—also apply to several other policy areas, but they are significantly more pronounced in the area of the environment and combine in ways that give environmental issues distinctive qualities. Few would argue that the environment is a private good. The atmosphere, for instance, is both non-rivalous and non-excludable. One person's use of air does not significantly affect another person's; and it is not possible to exclude persons or groups from access to the atmosphere.

Recent discussion has popularized the difference between "pure" public goods such as the atmosphere, with "impure" ones, for example forests and certain species (e.g., pandas, elephants). The former are labeled *common sink* resources,[32] while the latter are called *common pool* resources.[33] The difference is an important one, because the common pool resources can be allocated to individuals or groups, as a means to insure their sustainability.

Yet whether sink or pool, uncontrolled use of the environment leads to what Garret Hardin famously described as the "tragedy of the commons."[34] The metaphor refers to the results of unhindered private pursuit of a commonly-held resource—the oceans, forests, atmosphere, species of plants and animals. Excessive use leads to the depletion of the resource, unless regulated by government or, alternatively, privatized (in the case of common pool resources that can be allocated to individuals or groups).

Uncertainty refers to our lack of finite knowledge about the causes of many environmental problems and their long-term effects. For instance, intensive scientific investigation has not yet produced answers to the question of the exact role of anthropocentric factors compared to naturally occurring cycles in climate warming. Nor do we know what all the long-term effects of climate warming will be. In the area of biodiversity conservation, we do not know how many species the world has, and even for familiar charismatic species that are endangered (such as Siberian tigers and North Pacific sea lions) exactly what factors (and their interactions) explain population variability.

The third characteristic, irreversibility, provides drama to the environmental policy area. Endangered species easily may become extinct and cannot be recreated; heavily polluted rivers may turn black and die. These and other looming crises, the threats of an "end to nature," lend an urgency to environmental policy seen elsewhere only in issues of war and conflict. All three characteristics influence the coalition-building process.

6.2. The Policy Coalition Framework

To develop policies of environmental protection in liberal democracies requires formation of a majority, and that invariably means development of a coalition. Even in authoritarian systems, a majority of power holders needs to be convinced. Sabatier and Jenkins-Smith discuss the ways in which actors from a variety of institutions who share beliefs in a policy subsystem may form an "advocacy coalition."[35] As we saw in chapter 3, ENGOs are a prominent part and may lead such a coalition. However, the environmental policy subsystem (or issue network) includes opponents of change as well as advocates, and this makes change a function of competition within the subsystem as well as events outside it.

As James Q. Wilson notes,[36] the distribution of costs and benefits to advocates and opponents determines the strategy most likely to succeed in coalition formation. Here the nature of environmental problems and proposed resolution is especially relevant. The benefits of environmental policy change are likely to be broadly dispersed. Almost everyone benefits from cleaner air and water, from protection of threatened species and ecosystems, from curbing deforestation and desertification. At the local level, the benefits to environmental protection policy may be sufficiently large to stimulate organization of grassroots groups. Yet at the national level, benefits are diluted and do not form incentives for the organization of a policy change coalition. And as "pure public goods," most environmental benefits attract free riders.

In some environmental issue areas such as automobile emissions, the costs of controls (for example through eco-taxes) may be as dispersed as the benefits. In most cases, however, producers and other business groups bear the brunt of policy change, and costs are concentrated. This is a strong incentive for polluters and degraders to organize in opposition to change.

Under these circumstances of distributed benefits and concentrated costs, advocates of environmental policy change fight an uphill battle. Wilson suggests the necessity for either crises or entrepreneurs in the development of a winning majority coalition. The introduction to this volume gave examples of crises; each stimulated formation of advocacy coalitions, which in turn brought new policies and government responses into effect.

The environmental area also is a proving grounds for policy entrepreneurs, who would make political capital out of progressive environmental change. In the Exxon Valdez oil spill case, John Devens, the mayor of Valdez, led the state campaign for tough oil spill legislation and creation of an oversight council. One year later, he ran against the long-serving incumbent Don Young, who was Alaska's sole member of the U.S. House of Representatives, nearly defeating him. During the same time period, a member of Taiwan's legislature spearheaded opposition to development of a mammoth petrochemical complex and steel plant on the southwest coast of Taiwan, which would threaten the Qigu wetlands, a winter habitat for the globally endangered blackfaced spoonbill. The legislator, Su Huanzhi, organized a Homeland Protection Foundation, and held hearings and staged demonstrations throughout the affected communities and in front of the national legislature in Taipei. Then, he organized a 600 kilometer march, and after shaving his head bald in an act of Buddhist piety, led thousands of protesters. This stalled project development (which eventually collapsed due to change in business interests). Legislator Su, however, gained protected status for the spoonbill and the wetlands. He then used the successful campaign to win the Tainan county magistracy.[37] We suspect that many countries have similar stories about policy entrepreneurship.

6.3. Stages in Environmental Policy Development

The making of public policy is rarely a linear process. Instead, most scholars (and practitioners as well) consider it to be cyclical. As many as seven or eight categories can be identified in the policy cycle, but most authors include four common stages: policy formation, adoption, implementation, and impact (including evaluation and possibly change).[38]

Unlike our presentation of institutions, which is largely static, the stages of the policy cycle are dynamic. Although usually sequential, two or more stages may occur simultaneously. Thus, policy is being evaluated and often changed during the implementation process. We argue that the relevance of the comparative institutional differences discussed in earlier in this chapter is most apparent in the policy-making process, and demonstrate this through analysis of each stage.

Policy formation entails the identification of problems, and then placing them on an agenda for action (usually called a "government agenda"). Liberal democracies provide the best arena for the identification of environmental problems, because they give free reign to scientists and ENGOs, both of which monitor adverse ecosystem changes. A free press, also most likely to operate in liberal democracies, can place environmental issues on a public agenda instantly, making it difficult for decision-makers to ignore them. In contrast, authoritarian states can silence scientific reports, ban NGOs, and censor the press. These differences are glaring in countries of different political system type, but at similar levels of economic development, for example China and India.

The *policy adoption* stage displays clearly the difference between systems of concentrated and separated powers. Environmental policy becomes national law through the legislative process. In parliamentary systems such as those in most European nations, majority party leaders become prime ministers who drive policy proposals through the parliament to enactment, even in the face of substantial opposition. Presidential systems can operate with similar efficiency only when the different branches are under partisan control and when the legislature takes the lead of the president. In the U.S., unified governments in the post-World War II era have been no more frequent than divided governments. Separation of power systems encourage delay, stalemate and policy gridlock.

The *policy implementation* stage commonly is regarded as most critical to effective environmental protection outcomes. Universally, scholars note "implementation deficits" in environmental policy, which is a much over-used concept. Federal systems present challenges to environmental policy implementation, depending on both degree in centralization of the system and resources under central control that can be used to induce compliance by sub-national units with federal environmental mandates.

Most of the world's nations are unitary states, yet decentralization and devolution have become increasingly prevalent means to increase public involvement and to tailor policy to regional and local environmental conditions. Although federal systems give constitutional protection to the autonomy of sub-national units, political factors—local economic interests and partisan support bases—may make devolved or decentralized authorities equivalent to states and provinces in federal systems.

As Ribot notes, decentralization requires both the transfer of power to sub-national units and accountable representation. Summarizing the recent literature on decentralization experiments and their efficacy in environmental protection, he concludes:

> The potential of decentralization to be efficient and equitable depends on the creation of democratic local institutions with significant discretionary powers. But there are few cases where democratic institutions are being chosen *and* given discretionary powers. Ironically, a backlash is already forming against decentralizing powers over nature resources. Environmental agencies in Uganda, Ghana, Indonesia, Nicaragua, and elsewhere have argued that too much decentralization has caused damage or over-exploitation. These calls to recentralize control over nature resources are premature. Before decentralization can be judged, time is needed for them to be legislated, implemented, and take effect.[39]

What decentralization experiments lack, particularly in LDCs, is what federal systems of government provide: accountability at the sub-national level and discretionary power. This suggests then the advantage of federal systems in the implementation of environmental policy.

Finally, *policy impact and evaluation* distinguish clearly the separation of power from the concentrated power systems. By definition, separation of power systems permit, indeed encourage, penetration of the state through their multiple points of access for groups and non-governmental institutions. This stage also highlights differences between nations with independent court systems and those lacking autonomous judiciaries. The appeal to an independent tribunal is the most important recourse for those believing that policies are not having the results for which designed.

7. THE POLITICAL OPPORTUNITY STRUCTURE OF NATIONS

The nature and structure of political institutions in states influence the development of environmental interests, parties, and movements, as well as counter-movements in opposition to them. In this chapter we have focused on three institutional arrangements: federalism versus unitary state systems, concentrated versus dispersed powers, and constitutional versus authoritarian systems (with a special focus on court systems). These institutional arrangements are relatively invariant. In most cases, they were established at the formation or after significant adaptations to the nation-state, and they cannot be altered easily.

The "Third Wave" of democratization has brought some changes but their permanence in many cases is still questionable. Russia and most of the states of the former Soviet bloc have developed new political institutions, and Latin American countries have reinstated or vitalized dormant legislatures and electoral systems. These developments may yet open stable opportunities for the representation of environmental interests, independent judiciaries, and effective rule of law, but the record to-date is inconclusive.

Democratizing states and liberalizing economies do become more open to participation in international treaties and organizations. Unfortunately, for now, the resulting increases in trade and investment have tended to increase pollution problems. In the case of Mexico, for example, NAFTA's effects on institutions for environmental protection have not yet been positive. The North American Commission for Environmental Cooperation—created to implement the NAFTA environmental side agreements—does not compensate for cutbacks by the Mexican government in spending on environmental enforcement,[40] or the legacies of authoritarianism on the legislative and judicial systems.

Scholarly studies of differences in political institutions among nations suggest that corporatist systems correlate strongly with high environmental performance. Federalism, on the other hand, as well as dispersal of governmental powers tend to correlate poorly with strong environmental outcomes. "Consensual politics" correlates strongly with performance on environmental protection issues, but because several variables are related to consensus this hypothesis is less clear.

A review of the broader literature on movements and counter-movements places these institutional arrangements in a different light. This literature suggests that political institutions create incentives (or disincentives) for the organization of environmental interest groups, movements, and counter-movements. We note three generalizations that are relevant to the discussion in this chapter. The first pertains to both territorial distribution of authority and concentration of authority: "Movement-counter-movement conflicts are most likely to emerge and endure in states with divided governmental authority."[41] This hypothesis suggests that in unitary states, the state can make and implement environmental policy without serious challenge from internal opponents. In contrast, federal systems, such as the United States and Canada, allow environmentalists as well as anti-environmentalists opportunities for both support and opposition at different levels and branches of government.

A second proposition from the movement literature reinforces this point: "The availability of additional institutional venues for action encourages movements suffering defeats to shift targets and arenas to sustain themselves."[42] Federalism as well as dispersed authority systems (such as in separation-of-power systems) offer several venues for environmental action, which is increased by the existence of independent courts. This suggests that environmental groups and movements, as well as their opponents, may "venue-shop"—meaning they may seek out the most favorable institution to register protest and search for favorable action. The system overall is relatively more "open," providing general access to a number of groups, and this may increase the propensity for conflict. As we have seen, for LDCs, "venue-shopping" may involve appealing to international institutions for support in quarrels with domestic actors and institutions. Democratization and globalization increase the accessibility of these venues.

A third proposition refines these relationships between opportunities provided by structural arrangements and decisions leaders make on whether to mobilize and if so how. Based on analyses of anti-nuclear power movements in four post-industrial states, Kitschelt notes that there is a:

> curvilinear relationship between openness and movement mobilization, which shows that very closed regimes repress social movements, that very open and responsive ones assimilate them, and that moderately repressive ones allow for their broad articulation but do not accede readily to their demands.[43]

In sum, institutions matter. Values may underlie the original motivations of environmentalists, but political institutions establish the structures and relationships that influence how values become interests, interests become movements, movements become organizations, and organizations affect policies.

In chapter 5 we turn to specific cases of environmental policy-making, which illustrate the concepts and principles introduced in this section.

---

[1] See David Schoenbrod, *Power Without Responsibility: How Congress Abuses the People Through Delegation.* New Haven, CT: Yale University Press, 1993.

[2] Pastoral Juvenil Obrera, "The Struggle for Justice in the *Maquiladoras*: The Experience of the Autotrim Workers." In Timothy A. Wise, Hilda Salazar, and Laura Carlsen, eds., *Confronting Globalization: Economic Integration and Popular Resistance in Mexico.* Bloomfield, CT: Kumarian Press, Inc., 2003, 177.

[3] David Wheeler, "Beyond Pollution Havens," *Global Environmental Politics*, Vol. 2, No. 2 (May 2002), 5.

[4] Ibid.

[5] Article 26 of the 1982 Constitution reads:

> The state protects and improves the environment in which people live and the ecological environment. It prevents and controls pollution and other public hazards. The state organizes and encourages afforestation and the protection of forests. See: *Constitution of the People's Republic of China* (adopted by the Fifth Session of the Fifth National People's Congress on December 4, 1982).

[6] Barbara Lynch, "Development and Risk: Environmental Discourse and Danger in Dominican and Cuban Urban Watersheds." Iin Dimitris Stevis and Valerie J. Assetto, eds., *The International Political Economy of the Environment: Critical Perspectives*. Boulder and London: Lynne Reinner Publishers, 2001, 168-175.

[7] See for example Robinson, "Comparative Environmental Law: Evaluating How Legal Systems Address 'Sustainable Development'," *Environmental Policy and Law,* Vol. 27, no. 4 (1997), 338-45.

[8] For a discussion of the role of political institutions in general terms, see Gabriel Almond, G. Bingham Powell, Kaare Strom, and Russell Dalton, *Comparative Politics Today: A World View*, 7th edition. New York: Longman, 2000, 104-12.

[9] See Madison, "The Federalist No. 10," in Alexander Hamilton, James Madison and John Jay, *The Federalist Papers,* New York: Bantam Books, 1982, 42-49.

[10] Walter A. Rosenbaum. *Environmental Politics and Policy*, 5th edition. Washington, DC: Congressional Quarterly Press, 2002, 111.

[11] Arend Lijphart, *Democracy in Plural Societies: A Comparative Exploration*, New Haven and London: Yale University Press, 1977, 90.

[12] See John T. Rourke, *International Politics on the World Stage*, 10th ed. Boston, MA: McGraw-Hill, 2005, 202-03; and Albert Weale, Geoffrey Pridham, Michelle Cini, Dimitrios Konstadakopulos, Martin Porter, and Brendan Flynn, *Enviornmental Governance in Europe*. New York: Oxford Press, 2000, especially 86-107.

[13] Weale et al., 2000, 94.

[14] Louis D. Hayes, *Introduction to Japanese Politics,* 3rd edition. Armonk, NY: M.E. Sharpe, 2001, 147.

[15] Kenneth Lieberthal, *Governing China: From Revolution through Reform.* New York: W.W. Norton, 1995, 288.

[16] See, for example, the review by Shui-Yan Tang, Carlos Wing-Hung Lo, Kai-Chee Cheung and Jack Man-Keung Lo, "Institutional Constraints on Environmental Management in Urban China: Environmental Impact Assessment in Guangzhou and Shanghai," *China Quarterly*, Vol. 152 (December 1997), 863-74. The authors note:

> (L)ocal governments in China are confronted with difficulties in self-regulation. When a local government owns and sponsors many polluting industries or projects, it is difficult for it to monitor itself. When an environmental bureau is part of the local government, it is often difficult for the bureau to enforce environmental regulations on other government units that sponsor polluting factories and are of equal, and often higher, hierarchical ranking than the enforcing bureau itself. (864)

[17] Lennart J. Lundqvist, "Do Political Structures Matter in Environmental Politics? The Case of Air Pollution Control in Canada, Sweden, and the United States," in *American Behavioral Scientist*, 731-50.

[18] Kenneth K. Orie, "Constitutional Approaches to Sustainable Environmental management: Experience and Challenge," *Environmental Policy and Law*, Vol. 15, no. 1/2 (1995), 43-51.

[19] Uday Desai, *Environmental Politics and Policy in Industrialized Countries.* Cambridge, MA: MIT Press, 2002, 373-74.

[20] Elizabeth R. DeSombre, "Understanding United States Unilateralism: Domestic Sources of U.S. International Environmental Policy." In Regina S. Axelrod, David L. Downie, and Norman J. Vig, eds., *The Global Environment: Institutions, Law, and Policy.* Washington DC: CQ Press, 2005, 193-6.

[21] David Vogel, "Representing Diffuse Interests in Environmental Policymaking." In R. Kent Weaver and Bert Rockman, eds., *Do Institutions Matter? Government Capabilities in the United States and Abroad.* Washington, DC: Brookings Institution, 1993.
[22] Lawrence S. Rothenberg, *Environmental Choices: Policy Responses to Green Demands.* Washington, DC: CQ Press, 2002, 16-17.
[23] Desai, p. 378.
[24] Jon Shefner. "Coalitions and Clientilism in Mexico." *Theory and Society,* Vol. 30, No. 5 (October 2001): 593-628; Elizabeth Umlas. "Environmental Networking in Mexico: The Comite Nacional para la Defensa de los Chimalapas." *Latin American Research Review*, Vol. 33, No. 3 (1998): 161-89.
[25] Rosenbaum, 75.
[26] 16 U.S.C. Sec. 1540(g).
[27] Lettie M. Wenner, "Environmental Policy in the Courts." In Norman J. Vig and Michael E. Kraft, eds., *Environmental Policy in the 1990s*, 2nd edition. Washington, DC: Congressional Quarterly Press, 1994, 156.
[28] Tom Ginsburg, *Judicial Review in New Democracies: Constitutional Courts in Asian Cases.* New York: Cambridge University Press, 2003, 97.
[29] John S. Dryzek, David Downes, Christian Hunold, David Schlosberg with Hans-Kristian Hernes, *Green States and Social Movements.* New York: Oxford University Press, 2003, 39.
[30] Weale et al., 2000, 102.
[31] Neil Carter, *The Politics of the Environment.* Cambridge: Cambridge University Press, 2001, 162-68.
[32] See, for example, Albert Weale, *The New Politics of Pollution.* Manchester: Manchester University Press, 1992, 192-95.
[33] See Elinor Ostrom, *Governing the Commons.* Cambridge: Cambridge University Press, 1990, 30.
[34] Garrett Hardin, "The Tragedy of the Commons." *Science*, Vol. 162 (1968): 1243-48.
[35] See Paul A. Sabatier and Hank C. Jenkins-Smith, eds., *Policy Change and Learning: An Advocacy Coalition Approach.* Boulder, Colo: Westview, 1993.
[36] James Q. Wilson, *The Politics of Regulation.* New York: Basic Books, 1980.
[37] See Gerald A. McBeath and Tse-Kang Leng, *Governance of Biodiversity Conservation in China and Taiwan.* Northampton, MA: Edward Elgar Publishing, forthcoming 2006, ch. 8.
[38] For a standard interpretation of the policy cycle and stages, see James E. Anderson, *Public Policymaking,* 5th edition. Boston, MA: Houghton Mifflin Co., 2003.
[39] Jesse C. Ribot, *Democratic Decentralization of Natural Resources: Institutionalizing Popular Participation.* Washington, DC: World Resources Institute, 2002, 18-19.
[40] Kevin P. Gallagher, *Free Trade and the Environment: Mexico, NAFTA, and Beyond.* Stanford: Stanford Law and Politics, 2004, 84.
[41] David S. Meyer and Suzanne Staggenborg, "Movements, Countermovements, and the Structure of Political Opportunity," *American Journal of Sociology*, Volume 101, no. 6 (May 1996), 1637.
[42] Meyer and Staggenborg, 1648.
[43] Herbert P. Kitschelt, "Political Opportunity Structures and Political Protest: Anti-Nuclear Movements in Four Democracies," *British Journal of Political Science*, Vol. 16 (1986), 62. See also Dryzek et al., 2003, 100-01 for a treatment from the social movement perspective.

CHAPTER 5. NATIONAL CAPACITY TO PROTECT THE ENVIRONMENT

Differences in values, organizations, and institutions help explain why some nation-states have stringent environmental protection regimes, which are effectively implemented, and others do not. In this chapter we examine the variance in the "protective capacity" of nations, broadly meaning their ability to deliver good environmental outcomes to the citizenry. This discussion emphasizes to a greater extent than previously the differences between economically developed countries (EDCs) and lesser-developed nations (LDCs), because the distinction is so critical to the means available to nations to implement expensive policy choices.

The chapter begins with a definition of the concept of political capacity and then asks who influences environmental outcomes in nation-states. It considers the pre-requisites for capacity-building, and in particular economic, human, and political resources. It considers too the way in which these resources influence the national administration. The core of the chapter is description of three different types of capacity: environmental pioneers, environmental models, and the large class of "incapacitated" nation-states or "laggards" with respect to environmental protection. The chapter concludes with a discussion of policy learning, including learning about new policy instruments, through both vertical and horizontal diffusion.

1. THE CONCEPT OF CAPACITY

The term capacity is relatively new in the study of comparative politics; however the concept it taps—power—is as old as the disciplined study of politics itself. Capacity goes straight to the question of what individuals, groups, and forces are strong and which are weak in the life of nation-states. In its most general sense, capacity is the power to effect outcomes. To *effect* means to cause or bring about. Outcomes are actions or results. Capacity is thus the power to cause or bring about actions or results. In the context of environmental policy, capacity refers to the power of the nation-state to bring about results such as a lowered rate of carbon dioxide emissions.

Capacity is above all a capability or potential. It derives from the Latin *capacitas*, which means the ability to produce or an "ability, power, or propensity for some specified purpose, activity, or experience."[1] Capacity does not refer to a specific entity, such as a law, nor to a specific action, such as an executive veto. Instead, it is an ability that a person or institution possesses or that inheres in it. Depending on its specific form, this ability can be exercised by a robust leader, groups (such as environmental organizations), or institutions (for example, the Japanese Diet). However, capacity does not have to be exercised; it is a potential that can exist without being used. National capacity thus means the ability to determine or influence the decisions, actions, or behavior of state officials (as well as non-officials who can influence environmental results).

An important distinction is between two dimensions of capacity: dominance and influence. Both are forms of capacity in that they are ways of causing outcomes. Their difference is mainly one of degree: dominance implies a larger scope of capacity than influence. Formally, dominance is the maximum degree of political capacity. Dominant actors usually can obtain whatever they want from the government, for example business groups operating in modern Nigeria. Influence, on the other hand, is the capacity to effect outcomes indirectly or partially, without fully controlling them. For example, many of the non-governmental environmental organizations considered in this study have some say in pollution control policies of their nation-state but are unable to dictate the ultimate outcome. They have some access to state decision-makers but not complete control over their decisions.

2. WHO GOVERNS ENVIRONMENTAL OUTCOMES?

A leading question in environmental politics, indeed in any policy sector, is *who* has the capacity to effect environmental policy? Political capacity usually is wielded by those who hold official positions in government. Government officials possess legal powers or authority to take certain actions and make decisions, but the extent of this official legal power differs from one political system to another.

In democratic state systems, presidents and prime ministers enjoy statewide authority and as heads of government can set the agenda for action on national environmental issues such as water quality standards. However, in presidential and semi-presidential systems, executive authority is shared with legislative leaders, including chairs of environmental, budget-writing, and taxation committees. Too, in some democratic states judiciaries may have independent powers to resolve environmental controversies, as indicated by the U.S. Supreme Court's decision in 1978 to halt construction of the Tellico Dam to preserve the snail darter species. And, in federal systems, provincial ministers or state governors may strike more (or less) aggressive poses on environmental issues.

Authoritarian leaders such as the Chinese president, Nigerian or Burmese generals, even Gulf state sheiks typically possess more capacity to take official action on the environment than the leaders of most democracies. The lack of accountability of these leaders may even be an asset in their resolution of complex conflicts between pressures for economic development and sustainability. In short, the legal powers to effect environmental outcomes of individual leaders and members of various governmental institutions can be compared from one country to the next.

The capacity to determine or influence governmental actions is not confined to office-holders, whether presidents, party secretaries, or generals. In many cases, the business class (industrial sectors such as the energy and chemical industries or the private sector as a whole), environmental organizations, the media, or other groups that are formally separate from the state may have an extraordinary amount of control or influence over what the state does (or declines to do) regarding the environment. Because environmental problems tend to have complex origins, which are not easily understood by lay persons, groups possessing specialized knowledge—called *epistemic communities*—may have a great deal of influence over state policy, especially in drawing public attention to problems and in crafting both national and international mitigation strategies.

3. RESOURCES CRITICAL TO CAPACITY-BUILDING

No nation-state has achieved a condition of environmental sustainability, which means that each country confronts choices: how to use its scarce resources to build the capacity needed to develop and implement sustainability policies. Three sets of resources constrain (or facilitate) the development of capacity—economic, human, and political. In the next section we discuss administrative competence, the resource most proximal to effective implementation of environmental policy.

3.1. Economic Resources

We have emphasized throughout this study the correlation between level of economic development and environmental protection. Nations with a per capita GDP of less than $730 a year (or $2/day) will not have sufficient resources to clean up toxic wastes, polluted air, waters, and land. Such countries may lack even an integrated transportation and communications infrastructure to connect the population, and their extractive (tax collection) capacity is particularly weak. These are the poorest of the developing countries, and approximately 2 billion of the planet's people live in them. The economically developed countries, on the other hand, with per capita GDP of at least $10,000 have resources adequate for most forms of environmental remediation.

We do not posit a direct relationship between economic development and environmental protection policies (and outcomes) of nation-states. For example, countries at the same level of economic development may have different environmental outcomes, as we note below in the consideration of pioneers and models. Yet there is a correlation as noted in several previous discussions, and economic development is an essential pre-condition for the development of national capacity.

3.2. Human Resources

Broadly, human resources include educational attainment of national populations, as well as training in fields of applied technology and engineering (for example, training sufficient to operate equipment). Education and training constitute the human capital of the country, which varies greatly cross-nationally. Human resources also include collaborative abilities or social capital. Both human and social capital are strongly influenced by level of economic development in both positive and negative ways. Clearly, without budgetary resources, nations cannot establish educational institutions, hire teachers, and school their populations. Basic literacy and communication skills enable people to learn from multiple sources and share ideas with others. Environmental awareness cannot develop beyond isolated communities without populations educated at least through primary school, a condition not yet realized by the least economically developed nation-states. On the other hand, economic modernization is disruptive of traditional social structures and economic activities, including relationships of communities to ecosystems that have proven sustainable. In short, development may create new social capital needed to coordinate environmentally rational development policy while it destroys local social capital that has generated and preserved important information.

Because, on the national and international levels, solutions to most environmental problems require an understanding of scientific processes, and mitigating measures typically require technological means, the technological and scientific training of an element of the population is essential as well. A significant difference between EDCs and LDCs is the percentage of the population that has graduated from post-secondary institutions, which averages 25 percent in the former and less than 1 percent in the latter. Education in the environmental sciences is not available in most LDCs. The research and development (R&D) expenditures of governments and industry in EDCs also tower over those of LDCs, as do the development of specialized environmental research laboratories and technological institutes.

Finally, as Robert Putnam points out decisively in *Bowling Alone*,[2] the extent of social capital varies greatly across the globe and within each nation over time. Social capital describes the connections members of the nation have with others, their social networks, which establish relationships of trust that are critical to their working collaboratively together. Newly-developing nations are likely to have localized or parochial social capital, and to not have developed the degree of integration allowing citizens from different regions to work well with one another. Limitations of social capital obviously retard the development of participation in environmental organizations and in other civic organizations.

Here too development and modernization can cut two ways. Some LDCs, especially the highly vulnerable SIDS, have been active participants in international conferences, agreements and institutions on the environment, and have benefited from externally funded programs to install scientific monitoring equipment and train local stakeholders in its use and maintenance. When successful, these programs have significantly increased local capacity to generate useable data for planning and mitigation of the effects of climate change, sea level rise, biodiversity loss, and resource degradation.[3] But development guided by modernization theory (see chapter 2) still assumes the superiority and value neutrality of western, scientific methodologies in assessing risk and protecting environmental values. Other approaches to gathering knowledge about flora, fauna, ecosystems and climate may be lost or undervalued with modernization.

Recently, scholars and development professionals have come to recognize the additive value of scientific and traditional knowledge (also referred to with slightly different meanings as local, indigenous, or native knowledge, and civic science). Studies on climate change and biodiversity in particular have augmented empirical data with the observations and oral traditions of local populations. This has enhanced capacity in LDCs and EDCs, although the capacity to fully integrate the two types of knowledge to guide environmental policy-making is still underdeveloped. Where attempted, these approaches are closely associated with the participatory practices of resource management and environmental policy-making (discussed in chapters 2 and 3).[4]

3.3. Political Resources

A number of political factors influence capacity-building in nations, among which we consider stability, legitimacy, and transparency. A large difference between EDCs and LDCs is the greater political stability of the former. Many developing nations have had difficulty establishing a regime that lasts more than a few years, institutions have never become firmly entrenched, and routine practices have not been set. Under conditions of frequent regime change and political instability, it is virtually impossible to develop consistent and comprehensive environmental policies.

Second and related to stability is legitimacy of the state, the recognition of most people that it is a rightful political system. Many developing nations tend to lack legitimacy. Substantial minority populations may question whether the state should exist at all. Many in the population may doubt whether the regime should be kept. Under these circumstances, it is easy to transfer one's dissatisfaction with a government and its environmental policies (or lack thereof) to the regime or the nation as a whole.

A third important factor is the extent or prevalence of corruption within the nation-state, which is at least partly a product of lack in transparency of its political processes and institutions. This factor too is considerably more prominent in LDCs than in EDCs, and it extends far beyond the loss of already scant resources that are taken out of the country by the Marcos, Duvalier, Somoza, and other such families after they are deposed. The corruption is likely to penetrate deeply into the bureaucracy, especially where there is a lack of a strong legal system and other institutions to keep government employees in check. Because environmental protection policies typically affect business interests, they are particularly subject to the influence of corrupt dealings. In addition, successor regimes are left with both empty treasuries and high levels of external debt. They experience intense pressures to stimulate economic growth and return to creditworthiness as soon as possible. In the 1980s International Monetary Fund and World Bank structural adjustment programs encouraged highly indebted LDCs to concentrate on resource extraction with little regard for the environmental consequences. However, subsequent policies and programs by the World Bank have provided support for environmental capacity building.[5]

3.3.1. Administrative Competence in Environmental Policy-Making

Administrative competence refers to the policy-implementing organization of the nation-state with respect to environmental protection. It is influenced by the political institutions discussed in chapter 4, but is both analytically and practically distinct. We address two aspects of administrative competence. The first focuses on national environmental administrative institutions; the second examines the degree of strategic coordination in the environmental planning system.

3.3.2. National Environmental Institutions

The environment was a new issue in all countries, emerging only in the 1950s and 1960s. The issue developed in the context of industrialized and developed nation-states, which had well-established administrative structures organized by function. Initially, as environmental problems arose they were handled within relevant departments or ministries. Thus, health effects of air and water pollution occupied the attention of health ministries; problems of biodiversity loss and deforestation were relegated to agriculture, forestry, or land management ministries; and toxic waste disposal issues went to construction or defense ministries.

It was only in the late 1960s and early 1970s that certain nations established national environmental ministries. The first was in the United States, with the creation of the Environmental Protection Administration by executive order of the president in 1970. EPA is a cabinet-level agency whose chief administrator is appointed by the president. Currently, it has about 18,000 employees and an annual budget of approximately $8 billion. By the early 1970s, most industrialized countries had established environmental ministries or agencies.

In general, LDCs developed national environmental institutions a decade or so later than the EDCs, and a number of countries, particularly the smallest and poorest nations, have not established them yet. Thus the date national environmental institutions are established in countries is one measure of administrative competence.

A second measure is the extent to which the most important environmental functions are centralized in the national environmental agency or ministry. For example, the U.S. EPA's mandate includes regulation of air quality, water quality and protection, disposal of hazardous wastes, regulation of chemicals (including pesticides and radioactive wastes), as well as noise regulation. Yet some of these regulatory areas are shared with other departments; for instance, the disposal of hazardous wastes from military installations is primarily the responsibility of the U.S. Department of Defense, which has responsibility for clean-up of hazardous wastes on Formerly-Used Defense Sites (FUDS). In a number of countries, the national environmental ministry shares functions with as many as eight or nine different agencies, which is the case for the State Environmental Protection Administration (SEPA) of China.[6]

A third measure of competence is whether environmental institutions are established at sub-national levels of the political system. An example again is the United States where, by the early 1970s, each of the states had established departments of environmental protection or conservation. However, in the American example, one is less likely to find environmental bureaus in local governments. Instead, city and county health and land use departments are likely to have absorbed environmental protection functions. In this respect the American pattern is less advanced than that found in China, which has provincial and city/county environmental bureaus.

3.3.3. Strategic Environmental Planning

As Janicke notes, the stages of capacity-building also involve "mechanisms for environmental policy integration into different policy sectors; long-term environmental planning."[7] The issue is one of strategic coordination and development of integrated policy through what likely are dispersed institutions. Nations with coordinated strategic environmental planning systems are better able to address long-term goals, integrate environmental policy objectives into other policy areas such as transportation, and mobilize national resources to resolve environmental problems.

An innovative study by Janicke and Jorgens surveys environmental planning systems and actions in 17 OECD countries, and finds considerable variation across these developed nation-states.[8] The authors use three evaluative criteria, the first of which is the accuracy and relevance of environmental goals. They find that most of the national plans set a variety of general goals; few have quantitative targets with accurate time frames and detailed descriptions of measures to be taken. Most concrete is the Dutch National Environmental Policy Plan, followed by the South Korean Master Plan for the preservation of the Environment. Canada's Green Plan of 1990 offers a mix of qualitative and quantitative goals.

The second criterion is the degree of participation in and integration of the planning process. This refers to the extent to which environmental concerns are incorporated in other sectoral policies, as reflected in consultation and cooperation among relevant ministries. Reflecting the most intense cooperation was the Dutch practice. The third criterion is the extent of institutionalization of the green plan, for example enactment in a national environmental framework law or binding parliamentary decision, and establishment of a responsible institution with obligatory reports and budget documents. Janicke and Jorgens find that only five of the countries have national environmental laws incorporating green plans.

Three of the nations, the Netherlands, South Korea, and Sweden, have evaluated their planning regarding goal attainment. Although in none of the three countries had targets been met precisely, nevertheless failures were reported clearly, which led to a reformulation of policy. Based on the survey of the 17 countries, Janicke and Jorgens conclude that most green plans are pilot strategies with several deficits: goals tend to be unclear, targets qualitative and vague, and time frames absent. Although they believe that the green planning model is an innovation with the potential to increase political capacity at local, national, and global levels, presently it is at a preliminary stage.

4. COMPARATIVE ENVIRONMENTAL CAPACITY OF NATION-STATES

We explore national environmental capacity through an analysis of three different types of countries. The first and quite small group of states were environmental "pioneers," meaning they were the first to establish environmental agencies and broad-ranging environmental policies. The specific case presented is the United States. Although currently considered by some environmental policy scholars to be a "retrograde state," largely because of the Bush II Administration's withdrawal from the Kyoto Protocol, the United States nevertheless did play a leading role at the onset of the global environmental movement.

The second and equally small group of states have been "models" in the development of environmental capacity. Our representative state is Germany, considered by most environmental scholars to be the global leader in development of environmental institutions and policies, and we briefly refer to other states considered leaders in different aspects of environmental protection. The third and large group of states are "laggards," in the sense that they present cases of different kinds of "implementation deficits" which beset most nation-states in their attempts to develop comprehensive, integrated, and effective environmental policies. We have focused on three nations to represent this very large group of states—Russia, China, and Nigeria—and make brief reference to four transitional economies, Vietnam, Hungary, Poland, and the Czech Republic.

4.1. Pioneers

The United States has moved from a status as the world's leading nation in environmental protection to opponent of stringent environmental regulation, particularly in conservative Republican administrations following the election to the presidency of Ronald Reagan in 1980.[9] The United States entered the era of environmentalism with high per capita rates of fossil fuel combustion, serious air and water pollution, uncontrolled industrial waste disposal, waste water and waste management problems, rising rates of air pollution from vehicles, water pollution from agricultural soil erosion, and steady losses of natural wetlands and landscape.

The American system of policy-making, as compared to other post-industrial states, is quite fragmented—between national and sub-national governments, among and even between branches of government. The system offers countless opportunities for political mobilization and influence. The two-party system diffuses the impact of environmental change. Congress, the courts, local governments play stronger roles than elsewhere; the executive and bureaucracy are correspondingly weaker. A large number and diverse array of environmental organizations faces an equally diverse pattern of business groups. Some are conservative ideologically, a few benefit from environmental regulations but the most powerful opposition comes from economically weighty and high pollution sectors such as mining, the energy industry, logging, and agribusiness.

Andrews comments that environmental policy capacity in the U.S. has developed in patterns that are uneven, adversarial, and politically unstable. Although the EPA was the world's first environmental agency, and given sweeping regulatory powers and a large budget, it continues to lack an overall statutory mission. The U.S. has the most extensive scientific and technical capacity to support environmental policy-making, but support from proponents, while increasing after 1970, eroded somewhat because of economic fears during stagflation of the late 1970s and early 1980s.

U.S. capacity building is less influenced by international events than in other countries. American environmental policy is integrated only when strong leaders have taken the initiative; administratively, it is fragmented. The primary paradigm for pollution control policy has been national minimum standards and program requirements based on technological controls of industrial and municipal pollution sources and on substance-by-substance standards for hazardous contaminants. American "best practice" regulations have reduced pollution, but this system has been costly and inefficient. Alternatives promising lower costs—such as risk-based decision-making, pollution prevention, and market-oriented incentives—have been introduced only on a limited scale.[10] The most powerful market incentives in the U.S. are economic liability for pollution cleanup and the regulatory process itself.

Overall, U.S. environmental policies have corrected some of the most acute problems and slowed or worsened others. Scheberle cites these improvements:

> (N)ational airborne levels of lead, carbon monoxide, nitrogen dioxide, and ozone decreased between 1980 and 1999, sometimes significantly: 94 percent, 57 percent, 25 percent, and 20 percent, respectively. Approximately 30 percent less hazardous waste was generated and sent to treatment facilities, and roughly 60 percent of assessed U.S. streams, lakes, and estuaries could support their designated uses, such as fishing and swimming.[11]

The U.S. has pioneered many innovative concepts and instruments of environmental policy, for example the environmental impact statement, tradable emissions allowances, and citizen lawsuits to compel environmentally protective action. However, the basic structures of policy have not evolved toward pollution prevention and ecological modernization. Reforms remain experiments at the margin of regulatory mandates. Policies focus on pollution control, not prevention, with few serious restrictions on farms, land development, small businesses, local governments, and individual behavior such as vehicle use as compared to restrictions on large manufacturing plants.

4.2. Models

Germany entered the environmental era with geographic, political, and economic structures requiring greater environmental protection efforts than other countries.[12] It had a high population density and degree of industrialization, with environmentally problematical industries in a dense transportation network. Environmental problems of the 1970s and 1980s included local as well as transboundary air pollution, river and lake pollution and hazardous waste. Remaining problems include air pollution caused by private transport, illegal dumping, contaminated sites, growth of developed land, damage to trees, and soil and water pollution by agriculture.

In 1969 environmental policy became a national issue as a center-left coalition took power of government and adopted an official Government Declaration to solve environmental problems. This early action was not the consequence of public demands or interest group pressure, but was consistent with a long history of government regulation to protect humans against hazardous industrial activity. However, international developments influenced Germany, particularly those of the U.S. and the Council of Europe. German federal institutions were empowered, and three principles served as central guidelines: precaution, cooperation, and the polluter pays. Legislation enacted from the early to late 1970s comprehended most important areas in environmental protection. In 1990, the legislature adopted a program to reduce carbon dioxide emissions, and in 1994, environmental protection was written into the constitution as a goal of the state. Germany has one of the most complex legal systems for environmental protection in the world.

The German approach, a command-and-control type, consists of licensing and laying down standards, with strong administrative efforts at control in monitoring and enforcement. The government has stressed information and negotiation; and it has increased the number of voluntary industry-government agreements, which are a characteristic of a corporatist polity. Compared to other European states, Germany spends more on environmental protection. It uses emission-reducing technology most and has the strictest levels for air pollution, sulfur, and waste incineration plants. It was the first nation to stop production of CFCs, and had a "blue angel" (environmentally sound) product marketing system.

Germany has a broad network of well-established environmental policy institutions. The main federal actor is the Ministry for the Environment, Nature Conservation, and Nuclear Safety, formed after Chernobyl in 1986. The central government formulates, the Laender (states) and local authorities implement and enforce policy. The European Union has an impact on German environmental planning (as the EU is influenced by German environmental policy). Courts influence policy formulation and implementation. The number of environmental interest groups has increased steadily. The Green Party is firmly established in the party system, and in 1996 was the nation's third largest party. It joined the red-green governing coalition in 1998.

Janicke and Weidner note a steady increase in influence and competence of government and societal actors. The proportional voting system gives advantages to environmental interests, as does the type of federal structure Germany has developed. Participation rights were institutionalized early on, but "streamlining laws" have limited participation somewhat. Environmental interests have become integrated into parliamentary institutions. The state's informational capacity has grown as has the media's attention and focus on issues. Nevertheless, questions remain about whether institutions have sufficient capacity to realize long-term stabilization of the environmental situation (as represented in sustainable development), with problems in stabilization of land use, soil and ground water conservation, and reduction of material inputs into the production process.

Other countries have reputations as leaders in environmental policy planning, for example the Netherlands and Sweden. Japan has been described by some advocates of ecological modernization theory as an "ecological frontrunner nation," though critics believe institutional development in Japan is insufficient and the ENGO role too muted.[13]

4.3. Laggards

Most of the world's nations fail to address environmental problems effectively and often are called "laggards." We present three cases which illustrate different problem dimensions—China, Nigeria, and Russia—and make brief reference to Vietnam and three post-Soviet central European states.

4.3.1. China[14]

China is a communist country with a Leninist party that controls the state. Since 1978, however, marketizing reforms have reduced the party-state's control of the economy, which has somewhat weakened state capacity in environmental policy.[15]

China's population of 1.3 billion is the world's largest (about 22 percent), and this population is concentrated on about 12 percent of the land, resulting in one of the highest density ratios in Asia. Over-population and recent exceedingly rapid economic growth (in double digits for most of the 1990s) have depleted natural resources, causing deforestation, loss of farmland and grassland, and potable water sufficiency crises. Serious environmental problems also include air pollution (heavy sulfur dioxide emissions from coal-burning factories), acid rain, sandy desertification, pollution of most rivers, lakes and ponds near cities, copious toxic solid waste discharges, and urban noise.

Several historical and systemic weaknesses have contributed to China's environmental challenges[16]: (1) The Maoist-era command economy misallocated resources, because of high energy use per unit of GDP. (Marketizing reforms improved efficiency, and pollution increases now are lower than economic growth rates.) (2) Public ownership aggravated environmental damage. Lack of clearly defined ownership and property rights was the major cause of deforestation and depredation of grasslands. (3) The imperial tradition of personal rule has continued under communism. The legal system is poorly developed, and law enforcement lacks teeth. (4) Regulation is the major means of environmental protection and far more important than the law or economic instruments. Yet regulatory practice is influenced by traditional norms such as *Guanxi* (personal relationships). (5) Policy design focuses on government, not polluters, and does not follow optimal rules of resource utilization. (6) Finally, as the state devolved functions (including environmental protection) to provinces, special administrative regions, counties and municipalities, subnational governments focused on economic development; they had no incentives to provide training and funding for environmental protection bureaus (EPBs).

The central government is the primary actor in China's environmental policy development. The National Environmental Protection Agency was established in the early 1980s[17], reporting to the State Council, but has been influenced by the communist party (which controls personnel selection from top to bottom). The agency has broad powers in theory and a range of functions larger than that of the U.S. EPA, but it focuses on pollution control. The Chinese judicial system lacks independence. Mass media are largely state-controlled but increasingly give attention to environmental news. Pollution abatement essentially pits government departments against the industrial sector. State-owned enterprises (SOEs) have environmental offices to prevent pollution, but township and village enterprises (TVEs, the most rapidly growing section of the domestic economy since economic reforms began in 1978), China's current major polluters, are neither wholly public nor private and pay most attention to production costs. Although China has more than 2,000 environmental NGOs, the oldest and best funded are government-organized NGOs (or GONGOs) organized by the party and state; they have yet to play a significant role in pollution reduction. Since the mid-1990s, international NGOs such as Greenpeace, WWF, Conservation International, and The Nature Conservancy have set up shop in China, and they have had a marginal impact in reducing deforestation and protecting endangered species.

The environment figures increasingly in China's political development. Since 1978, the state has enacted a fairly comprehensive system of environmental laws and added articles on environmental protection to the constitution. The tally in 2005 includes 20 environmental laws, close to 50 administrative orders issued by the State Council, over 170 rules and regulations issued by SEPA individually or jointly with other agencies, in excess of 2,000 local laws and regulations, and an equally large number of environmental standards. Too, China has entered into more than 50 multilateral environmental agreements and signed at least 40 bilateral accords with different countries.[18] China's college and universities train students into environmental careers and raise awareness of environmental issues. These are promising developments.

Estimates of China's environmental situation in the early 21st century vary by observer. Progress has relied on growing environmental consciousness of the authoritarian party-state elite and significant assistance from foreign funding agencies (World Bank, Global Environment Facility), foreign governments, and international NGOs. Together they are estimated to comprise the lion's share of China's spending on environmental protection. Future environmental reform is dependent on continued economic growth, economic restructuring of SOEs and TVEs, and increased transparency (to reduce corruption and allow development of popular awareness of environmental problems and pressures for change).

As an aside, China is not the only transitional economy still operating under Leninist party-state rules. Vietnam also experiences large environmental planning and implementation difficulties, exacerbated by the ravages of war for most of the period between 1945 and 1975. Vietnam began marketizing reforms later than China, in 1986. Since then it too has privatized state-owned enterprises and decollectivized agriculture. The state's focus on economic development did not shift to mitigation of environmental burdens until the early 1990s, and the apparatus of legal instruments and implementing agencies remains incomplete. Unlike China, whose cautious political liberalization permitted organization of ENGOs and international NGOs in the mid-1990s, as of 2002 Vietnam had only one ENGO (the Vietnam Association for the Protection of Nature and Environment), with strong linkages to the government.[19]

4.3.2. Nigeria[20]

Nigeria is Africa's most populous country. Intensive agricultural practices and industrialization have worsened the environment. For example, chemical and cement industry pollution coat the countryside with dust. Petroleum production—the motor of the economy since the 1970s—has devastated oil field areas. Wood is the dominant domestic fuel source, and indoor pollution is severe. Excessive wood cutting, infrastructure development, and industrial siting have nearly deforested the state. Heavy use of mineral fertilizers pollutes rural waters; garbage covers urban areas.

Since independence from Britain in 1960, most Nigerian rulers have been of the military. Although Nigeria is a federal state, legislative powers of the state have been usurped by the center. The environment is not mentioned in the constitution. Although Nigeria's oil wealth has quickened economic change, it also has corrupted political institutions and processes.

Environmental policy responds to internal pressures as well as to the ambition of leaders who seek to play a leading role in the African continent. The 1989 National Policy on the Environment was Africa's first, and Nigeria has signed 30 international environmental conventions (yet few have been ratified and even fewer implemented). Although the government published national development plans soon after independence, it did not establish a federal environmental protection agency (FEPA) until 1988 and did so then only because of the Koko episode (Italy's illegal dumping of 4,000 tons of industrial wastes). Prior to this, the only regularly implemented program was an environmental sanitation drill, which compelled citizens to clean their homes and working environments on the last Saturday of every month, from 7 to 10 AM.

Legislation enacted after 1989 regulated sectors such as atmospheric pollution and hazardous wastes. The most comprehensive law is the Environmental Impact Assessment Decree of 1992. In general, policy is fragmented and uncoordinated; prevention, regulatory, and penal functions of state and federal governments are scattered among many different agencies. A new policy paradigm, developed with World Bank assistance, promoted an "anticipate-and-prevent strategy" for sustainable development; it emphasized strong support for incentives and other fiscal tools. However, successive governments enact stringent environmental laws with no expectation that they will be enforced effectively, due to limited human and material resource capacities. Thus, a "persuasion through dialogue" policy was introduced in the early 1990s; it asks industries to report regularly on voluntary industrial pollution abatement measures. The government claims that its "polluter pays" principle is its underlying environmental policy, but this policy is ignored with respect to oil multinationals and domestic conglomerates.

A large number of central environmental institutions are responsible for policy development and implementation. This appears to be the product of: (1) elite preference for technical solutions to political problems (a practice dating from the colonial era), and (2) elite inclinations to create institutions to satisfy material ambitions of egoistic individuals. The state lacks sufficient power to corral strong (and errant) industrial enterprises or enlist the public in environmental decision-making. Courts play no role in environmental management.

Environmental NGOs have some influence, but Nigeria lacks a robust civil society able to monitor the state and curb powerful economic interests. It also lacks a Green party. One emerging environmental pressure group is the occasionally violent Movement for the Survival of the Ogoni People, formed to seek compensation for oil exploitation of this region (see chapter 3). Nigeria has more than two dozen urban-based environmental NGOs, partly penetrated by the elite and industry or influenced by multilateral donor agencies. Two women's environmental groups formed in the mid-1990s. A previous president's wife founded a "better life movement," and it has tackled barriers to women's development in rural areas. Nigeria does have a long tradition of media involvement in environmental issues, notwithstanding strict press laws.

Capacity-building activities in Nigeria include some efforts in training of environmental managers, but to date have not stressed citizen involvement. The World Bank estimates that environmental deterioration causes health risks to 50 million Nigerians and expenditures of 3-20 percent of GDP. The state allocates 1 percent of the federal budget to ameliorate environmental degradation caused by natural disasters. Most funding for federal and state environmental activities comes from international donor organizations.

Nigeria's capacity for environmental policy is limited by economic and political instability, dependency, and the overly centralized and politicized bureaucracy. The state relies on imported technological expertise for industrial development; it is highly porous to multinational corporations. Although global financial institutions have provided significant support to Nigerian development, this support (especially structural adjustment programs of the World Bank and IMF) has eroded the administrative basis for environmental policy.

4.3.3. Russia[21]

Russia's environmental problems largely result from the economic policy of a centralized planned economy. Low energy costs promoted waste; the centralized pricing system rewarded environmentally harmful behavior. Obsolete and exhausted production plants that inefficiently consumed raw materials and energy and had high repair costs created a huge environmental burden. Industrial activity created most of Russia's air pollution problems; the recent reduction in emissions is due not to environmental protection but to the decline of industrial production. There is a very low level of water conservation. Inland waters are polluted by oil and organic materials. Huge toxic waste deposits, nuclear waste and safety are other environmental problems.

Environmental problems prompted the Soviet state to define environmental protection as a national goal in 1972. The system at that time emphasized clean-up measures and was fragmented with overlapping areas of competence. Not until 1988 was a state organization given responsibility for the entire environmental policy—the State Committee for Environmental Protection. The share of environmental investment as a percentage of total investment in the 1970s and 1980s was 1.5 percent, and most of this was directed to water conservation, with smaller portions allocated to air quality improvement and land conservation. Environmental investment fell significantly after the dissolution of the Soviet Union.

Current Russian central environmental institutions resemble those of the Soviet era. The Ministry for Environmental Protection and Natural Resources is responsible for policy development, coordination, and implementation. Gorbachev's glasnost and perestroika policies stimulated emergence of environmental NGOs. First came small-scale, local organizations to solve specific problems. Then, regional and national organizations developed, conducting public demonstrations against environmental hazards such as construction of the Volga-Don canal. A Russian Green Party formed in 1991 but it lacks political weight. By 1993, about 800 NGOs claimed notice, and they participated in campaigns and parliamentary elections. Greenpeace and WWF, among other international NGOs, now have Russian representatives; yet domestic green organizations lack finance, up-to-date equipment, and organization. The lifting of press restrictions led to greater but relatively specialized coverage of environmental issues, but under President Vladimir Putin the state has reasserted some of its former control. There are few green business enterprises.

Notwithstanding the relatively highly developed institutional structure of environmental authorities and formal/legal administrative instruments, environmental policies are poorly implemented. As compared to western nations, Russia lacks a democratic tradition based on the rule of law, and this is an impediment to implementation of environmental regulations. For example, since 1991 the state has levied charges on industry for pollution and for use of natural resources. The charges are regionally differentiated and oriented to hazards posed by pollution, but are set at low rates. A new environmental protection law in 1994, designed to adapt environmental legislation to the federal structure of the constitution, enables greater public participation on environmentally important issues, but the law has loopholes regarding waste disposal, safety of industrial technology, and nuclear safety.

About half of environmental investment in the 1990s came from partially privatized companies, state corporations, and the communal sector. The financing scheme has few sanctions against offenders and imposes mild punishments. State budget deficits further reduce state support for environmental protection. International cooperation, however, has been a positive force in capacity-building. For example, 17 German-Russian projects have improved the environmental soundness of oil and gas extraction and created an environmental monitoring system. The U.S. and World Bank have made some contributions too.

In the 1970s and 1980s, the policy approach consisted of end-of-pipe, high chimney strategies to deal with pollution. In the 1990s, the paradigm shifted to precautionary and preventive methods. Altogether, Russia's attempts at ecological modernization at best have had modest success. The main restriction to capacity-building is the deep, prolonged crisis in the national economy. Most observers believe the state sorely needs investment for technological modernization and environmentally clean production processes.

To Potravny and Weiszenaburger, the Russian case demonstrates that there are two indispensable preconditions for a successful environmental policy: (1) a stable, functioning political system able to influence the most significant developments in the country; and (2) a stable, functioning economic system to provide a material foundation and the minimum technical possibilities for environmental protection.

Three post-communist nations of Central Europe—Hungary, Poland, and the Czech Republic—face implementation deficits too. For 40 years, these states had centrally-planned economies and an inefficient system of energy utilization, which made them among Europe's most polluted states. Under communist rule they developed some environmental legislation. The democratizing reforms have increased governmental transparency and provided a nourishing environment for growth of ENGOs. Although civil society in the three states is less robust than in most western European nations, it is a definite contrast with the communist past. What most distinguishes environmental practice in the transitional states, in comparison with Russia, is requirements of the European Union. Hungary, Poland, and the Czech Republic were among the 10 states joining the EU in 2004. Each was required to conform to environmental management and emission targets as a condition of entry into the Union.[22] Moreover, since the late 1990s, the

EU has provided assistance to all three states under the Poland and Hungary Assistance to the Restructuring of the Economy (PHARE) program, and the EU linkage has been helpful in opening up the spigot of environmental planning and project assistance from other international donors. Thus, the prospects for these three states are significantly better than for Russia.[23]

4.4. Exceptions: Capacity Building in SIDS

Small island developing states (SIDS) present a special but not unique challenge to building capacity for environmental policy. SIDS of the Commonwealth Caribbean (former British colonies such as Barbados, St. Lucia and Grenada), for example, have many of the attributes of "pioneers" and "models." Most are democratic, middle income countries with high literacy rates, and an organized, politically aware citizenry. But even upper-middle income SIDS like Barbados, because of their economic resource limitations and small populations, may be poor in the human and political resources that provide sufficient national capacity to protect the environment. Their dependent economies make them vulnerable to external change. As traditional agricultural exports such as sugarcane and bananas decline they are replaced by tourism and light manufacturing, and geography leaves them unusually vulnerable to environmental threats that can jeopardize investment in these industries.[24]

Small pools of human resources, however, do not prevent duplication and overlap of jurisdiction in environmental planning and resource management. In Grenada during the 1960s and 1970s, for example, land use planning and zoning suffered from jurisdictional disputes between municipal and national authorities. Economic, development, flood control, forestry, fisheries and wildlife conservation responsibilities were fragmented among at least ten ministries and departments. Coordination among organizations was weak and national government approaches to environmental policy tended to be ad hoc and short term.[25]

Since the 1990s, however, SIDS have received assistance from bilateral aid agencies, international organizations and international ENGOs to build capacity for environmental protection. The effects have been inconsistent but not insignificant. ENGOs active in environmental education and training in SIDS include the Island Resources Foundation (IRF), CANARI (see chapter 3), and RARE. The World Bank has made capacity building for LDCs a larger part of its lending portfolio, using it to support the environmental conditionalities attached to its development funding. And official development assistance agencies from Great Britain, France, Germany, Canada, and the European Union have sponsored programs to establish protected areas (see chapter 6), parks, conservation plans, sustainable resource development strategies that include the training of local personnel and the provision of equipment and salaries for the expansion of agencies and the development of parastatal organizations and public-private partnerships.

In Grenada in the late 1990s assistance by the British Department for International Development successfully enhanced the capacity of the Forestry Department, contributed to the creation of a new national park and a protected area, the development of a sustainable forestry policy. The department received much needed equipment and forestry officers were trained in British universities. The department expanded its jurisdiction to include national parks and it has become the main locus of environmental conservation policy implementation in the country.[26]

Such successes, however, may be exceptional. Several Caribbean countries have attempted to establish sustainable development councils to overcome capacity problems. The councils are meant to provide a balanced approach to sustainable development by convening state, NGO and private sector representatives to advise the government on development policy. In the Eastern Caribbean six were attempted and only one survives. In most cases the councils were unable to secure the necessary recognition and support of government planning and resource management agencies which were closely linked to critical industries such as construction, manufacturing, export agriculture, and conventional (as opposed to eco-) tourism.[27] In government, environmental portfolios have a tendency to migrate among the ministries of public health, agriculture, planning or development. Typically, the functions, influence and budgets of environmental departments are subordinated to central missions of the ministries to which they are assigned (economic development, resource extraction, disease control, etc.) and their effectiveness rests on their ability to attract and maintain external support.[28]

5. GLOBAL ENVIRONMENTAL POLICY LEARNING

The case studies[29] show a close relationship between per capita GDP and environmental indicators. Wealthier countries both need and can afford more environmental protection. Although the relationship between affluence and environmentalism is significant, it also is contradictory: economic development leads to both improvements and deterioration in environmental quality, with significant differences in environmental area. For example, reductions in sulfur dioxide emissions and the extension of sewage systems parallel growth in GDP. But carbon dioxide emissions and fertilizer consumption increase up to a certain per capita GDP and then stabilize. Road traffic emissions are higher in rich states. The volume of waste generated also rises in rich states. There is an increasing accumulation of pollutants in rich states (which are older, industrial societies) and less biodiversity. The developing nations show general environmental deterioration—high air/water pollution rates, problems with particulates and both desertification and deforestation. However, the developing nations start from a lower base in per capita emissions rates and have less accumulated pollution.

In general, capacity-building begins with establishment of specialized government institutions, which occurred at different times comparatively and involved several processes of diffusion—both horizontal and vertical. Environmental pioneers such as the United States and Sweden initiated policy innovations such as the U.S. National Environmental Protection Act and institution building (establishment of the EPA). These ideas—to create specialized regulatory and advisory bodies, comprehensive framework laws, and even constitutional principles—then were diffused from the pioneers to most industrialized and to some developing countries. Also, pioneers diffused environmental instruments, such as Environmental Impact Statements (EIS), liability rules, emissions management, and eco-labeling. Overall, command-and-control measures appear to have been diffused earlier and more quickly than softer approaches such as participatory management and negotiated regulation instruments.

At the outset, the UN Conference on the Human Environment (Stockholm, 1972) was the most important institutional mechanism for vertical diffusion to advanced industrial countries. For developing countries, the critical conference was the UNCED in Rio de Janeiro (1992). The first wave for developing countries paralleled the second wave for post-industrial states (consisting of long-term goal setting, intersectoral integration, and cooperative target group policy). While the United Nations environmental conferences played critical roles in vertical diffusion, global financial organizations such as the World Bank, and multinational corporations, also became agents of diffusion.

Both pioneers and models have played critical roles in global environmental capacity-building, but the process has been inconsistent. Some innovators took the global spotlight and then retreated, for example Great Britain, the U.S., and Japan. Others rose, reached a plateau, and then declined, for example Germany. Sweden and the Netherlands have been consistent leaders. Retrogression may be explained by loss of national capacity, structural crises in the nation's economy inciting domestic opposition to environmental reform, and under-utilization of existing institutional, economic, or informational capacities.

In most countries, environmental NGOs have played influential roles in capacity-building. Environmental NGOs in the U.S., Netherlands, and Sweden are larger than political parties. Mass media have been a relevant factor in most countries, and are interrelated with the rise of environmental NGOs. Where systemic conditions for environmental policy-making are poor, NGOs may be the main driving force, as for example in China and Russia.

NGO influence in the early stages relies on the use of unconventional methods. Later, as the influence of NGOs increases, Janicke & Weidner hypothesize that they become more cooperative. The most important precondition for strengthening the NGO role is the presence of democratic rules providing for formation of autonomous social organizations. As noted in chapter 3, this varies greatly cross-nationally. Too, legal provisions for participation, rights to information, and civil court actions are important. But the role of courts varies cross-nationally. Trade unions tend not to be proponents of environmental protection. Green parties are obvious strong proponents, but they are not a necessary condition for environmental policy innovation. Green businesses (committed to environmentally sound management practices and sustainable development) are increasing in number.

What, then, are the essential preconditions for environmental capacity-building? Most basic are political and economic stability. A second factor is internal integration of relevant government environmental activities (with most countries experiencing difficulty integrating environmental and transport policy). A third factor is the integration of forces external to the state: business and then social organizations. States need to win the voluntary cooperation of polluters who are the target group, which is far easier for countries with a corporatist pattern of state-society relations.

It is no surprise that the nation-states praised for environmental leadership—Germany, Sweden, the Netherlands, even (and more recently Japan)—all have state-society relationships including aspects of corporatism. In the United States, pluralism and fragmentation of political institutions means that environmental NGOs lack influence over system outputs (implementation and enforcement of environmental laws). Because environmental issues are complex and technical, the epistemic community may play an important role as it commands available knowledge about problems and options. The media are vital in creation of public awareness. Conditioning factors include existence of environmental crises, as it is easier to build capacity in response to imminent catastrophes than to less spectacular, gradual forms of ecological degradation. Finally, the overall economic situation affects opportunities of environmental proponents. Economic performance decides whether economic instruments will be effective or not.

Capacity-building depends on material and human resources. It needs to be kept in mind, though, that the character of the environmental problem is a main factor influencing outcomes. Some problems such as traffic emissions, waste production, soil contamination and extensive land use have not undergone improvement in average countries. Others problems, such as sulfur dioxide emissions and municipal sewage have improved nearly everywhere. The nature of the problem invariably influences responses of institutions. We see this more clearly in chapter 6, which discusses national responses to global environmental problems.

---

[1] Lesley Brown, editor, *The New Shorter Oxford English Dictionary*. Oxford: Clarendon Press, 1993, 332.
[2] Robert Putnam, *Bowling Alone*. New York: Simon & Schuster, 2000.
[3] For example, the UN and Global Environment Facility funded the Caribbean Program for Adaptation to Global Climate Change in 1995; see, Bruce Potter, *CARICOM Global Environment Facility Project: Planning for Adaptation to Climate Change*. Island Resource Foundation, 1996.
[4] Billy R. DeWalt. "Combining Indigenous and Scientific Knowledge to Improve Agriculture and Natural Resource Management in Latin America," in Francisco J. Pichón, Jorge Uquillas, and John Frechione, eds., *Traditional and Modern Natural Resource Management in Latin America*. Pittsburgh: University of Pittsburgh Press, 1999, 101-24; and Karen Bäckstrand. "Civic Science for Sustainability: Reframing the Role of Experts, Policy-Makers and Citizens in Environmental Governance," *Global Environmental Politics*, Vol. 3, No. 2 (November 2003): 24-41.
[5] David Reed, "The Environmental Legacy of Bretton Woods: The World Bank," in Oran R. Young, ed., *Global Governance: Drawing Insights from the Environmental Experience*, Cambridge, MA and London: The MIT Press, 1997.
[6] Whether concentrated in one ministry, shared with several, or assigned to an operating agency, administrative placement influences coordination. This is especially problematical in implementing sustainable development programs. For a review of the coordinaton challenge, see Phillip J., Cooper and Claudia M. Vargas, *Imiplementing Sustainable Development*. Lanham, MD: Rowman & Littlefield, 2004, 235-43.
[7] Martin Janicke, "The Political System's Capacity for Environmental Policy: The Framework for Comparison," in Helmut Weidner and Martin Janicke, eds, *Capacity Building in National Environmental Policy*. Berlin: Springer, 2002, 13.
[8] Martin Janicke and Helge Jorgens, "National Environmental Policy Planning in OECD Countries: Preliminary Lessons from Cross-National Comparison," *Environmental Politics*, Vol. 7, No. 2 (summer 1998): 27-54.
[9] The treatment of the United States follows the discussion by Richard N.L. Andrews in Janicke and Weidner, eds, *National Environmental Policies: A Comparative Study of Capacity-Building*. Berlin: Springer, 1997), 1-24.
[10] For an additional review of U.S. regulatory inflexibility, see Daniel J. Fiorino, "Flexibility," in Robert F. Durant, Daniel J. Fiorina, and Rosemary O'Leary, eds., *Environmental Governance Reconsidered: Challenges, Choices, and Opportunities*. Cambridge, MA: The MIT Press, 2004, 393-426.
[11] Denise Scheberle, "Devolution," in Durant, Fiorina, and O'Leary, 2004, 361.
[12] This section follows closely the treatment of Germany in Janicke and Weidner, eds, 1997, 133-55. See also Helmut Weidner, "Environmental Policy and Politics in Germany," in *Environmental Politics and Policy in Industrialized Countries*, Uday Desai, ed. Cambridge, MA: The MIT Press, 2002, 149-202.
[13] Andrea Revell, "Is Japan an Ecological Frontrunner Nation?" *Environmental Politics*, Vol. 12, No. 4 (Winter 2003): 24-48.
[14] This section follows the treatment of China by Yu-shi Mao in Janicke and Weidner, Editors, 1997, 237-56, and is updated to 2005.
[15] For a thorough discussion of China's political and economic systems, see Kenneth Lieberthal, *Governing China: From Revolution Through Reform*. New York: W. W. Norton, 1995, and Tony Saich, *Governance and Politics of China*, 2nd edition. New York: Palgrave, 2004.

[16] For recent treatments of China's environmental challenges, see, among others, Judith Shapiro, *Mao's War Against Nature*. Cambridge: Cambridge University Press, 2001; Elizabeth Economy, *The River Runs Black*. Ithaca, NY: Cornell University Press, 2004; and Kristen A. Day, ed., *China's Environment and the Challenge of Sustainable Development* Armonk, NY: M. E. Sharpe, 2005.
[17] In 1998, the name of the agency was changed to State Environmental Protection Administration (SEPA) and it was raised to ministerial status.
[18] Bie Tao, "Environmental Law System in China," *China Daily*, October 2, 2005, 8. See also Jonathan Herrington, "'Panda Diplomacy': State Environmentalism, International Relations and Chinese Foreign Policy," in Paul G. Harris, ed., *Confronting Environmental Change in East and Southeast Asia*. Tokyo: United Nations Press, 2005, 102-18.
[19] See Le Thac Can, "Environmental Capacity Building in Vietnam," in Helmut Weidner and Martin Janicke, eds., *Capacity Building in National Environmental Policy*. Berlin: Springer-Verlag, 2002, 392-407; and Thomas O. Sikor and Dara O'Rourke, "Economics and Environmental Dynamics of Reform in Vietnam," *Asian Survey*, Vol. XXXVI, no. 6 (June 1996): 601-17.
[20] This section summarizes the presentation of Nigeria by Fatai Kayode Salau in Janicke and Weidner, 1997, 257-78. See also Olusegun Areola, "Comparative Environmental Issues and Policies in Nigeria," in Uday Desai, ed., *Ecological Policy and Politics in Developing Countries*. Albany, NY: State University of New York Press, 1998, 229-66.
[21] This section summarizes the presentation on Russia by Ivan Potravny and Ulrich Weiszenburger, in Janicke and Weidner, 1997, 279-98.
[22] Each state was required to successfully integrate the *acquis communautaire,* including that on the environment, as part of the domestic legislation of the country.
[23] See Adam Fagin, "The Czech Republic," in Weidner & Janicke, 2002, 177-200; Joanne Caddy and Anna Vari, "Hungary," in Weidner & Janicke, 2002, 219-38; and Magnus Andersson, "Environmental Policy in Poland," in Weidner & Janicke, 2002, 347-74.
[24] John W. Ashe. *Tourism investment as a tool for development and poverty reduction: the experience of Small Island Developing States (SIDs)*. Barbados: the Commonwealth Finance Ministers Meeting, 2005 http://www.sidsnet.org/docshare/tourism/20051012163606_tourism-investment-and-SIDS_Ashe.pdf.
[25] *Grenada Environmental Profile*, St. Michael, Barbados: Caribbean Conservation Association, 1991, 224-9.
[26] Stephen Bass. *Participation in the Caribbean: A review of Grenada's forest policy process*. London: International Institute for Environment and Development, 2000.
[27] Jonathan Rosenberg and Linus Spencer Thomas. "Participating or Just Talking? Sustainable Development Councils and the Implementation of Agenda 21." *Global Environmental Politics*, vol. 5, no. 2 (May 2005): 59-87.
[28] Rosenberg and Thomas (2005); and Jonathan Rosenberg, "The Local, the National, and the International in Sustainable Development Policy: Two Cases from the Eastern Caribbean," *Journal of International Wildlife Law and Policy*, forthcoming.
[29] This section summarizes the conclusion to Janicke and Weidner, 1997, 299-314.

CHAPTER 6. NATIONAL RESPONSES TO GLOBAL ENVIRONMENTAL PROBLEMS

1. GLOBAL ENVIRONMENTAL POLICY ISSUES

Several of the environmental problems we have discussed are primarily domestic in origin. They arise from accidents of geographic location and natural resource endowments—such as sufficiency of water—and are exacerbated by population pressures and economic development. Others, such as severe drought in regions of the world, and depletion of water resources may have more complex and distant origins and have prompted tensions between nations, seen particularly in Africa, Asia and parts of Latin America. The global and the national have become increasingly difficult to disentangle. Management of international river systems by upstream countries—such as the Colorado flowing from the U.S. to Mexico—has long been a source of international disagreements over the volume and quality of water that reaches the downstream countries. Nevertheless, water sufficiency for the present remains a domestic concern; most aspects of land and air pollution and refuse disposal remain national and local responsibilities, and many endangered species (and their habitats) are contained within national boundaries.

But a suite of issues including climate change, biodiversity loss, deforestation, desertification, ocean pollution, transboundary air pollution (including acid rain), and toxic waste disposal have become international problems, indeed crises, in the last two decades. They are considered global because they originate in more than one nation-state, and their effects appear in many states and, for the case of climate warming, in all. Single nation-states may take actions to ameliorate the effects of these issues, but cannot resolve the root problems. As a result, action through the development of international organizations, international NGOs, and international conventions and treaties is necessary.

In this chapter, our subject is the ways in which nation-states respond to global environmental problems. We are interested to discern:

- How consciousness has developed concerning the global issues, and particularly the response of scientists and civic organizations to it;
- How government policy has evolved concerning the issue;

- What role the nation-state has played in the formation of international regimes, conventions, and laws; and
- What if any ameliorative or adaptive actions the nation-state has taken.

We approach these questions through presentation of two cases in global environmental policy-making: climate change and biodiversity loss. In both cases, we discuss briefly the scientific background to the issue, relevant international conventions, and then we contrast the way economically developed and developing nations have responded. These discussions pull together analytical threads from chapter 2 (state-society relations), chapter 3 (political processes and organizations), chapter 4 (political institutions), and chapter 5 (national capacity) to suggest ways in which values, attitudes, institutions, and levels of development have determined the types and efficacy of responses by nation-states.

2. CLIMATE CHANGE

2.1. Scientific Evidence

Increasingly, the facts of global climate change seem incontrovertible. Careful scientific measurements establish that carbon dioxide content of the atmosphere, largely a consequence of the burning of fossil fuels such as coal, wood, and oil and gas, has increased nearly 20 percent since the start of the industrial revolution, a trend which follows the upward growth in global population. Observations from meteorological stations across the far North show increases in annual mean temperatures of up to 1 degree centigrade over the last generation. Observed impacts in the North (impacts are greatest in the world's polar regions) include: melting of glaciers, shrinking of sea ice extent in the Bering Sea, and thawing of permafrost in Alaska and Siberia.[1] Climate change effects in temperate and tropical regions of the world have been less extreme. Nevertheless, island states have been particularly susceptible to small rises in sea water levels and more intense storms.

Future consequences of climate change are expected to be more dramatic and disruptive to human life than the near-term observed results. For instance, rising sea levels, erosion, and storm surges may necessitate extensive relocation of coastal communities, huge expenses for reconstruction of infrastructure, and increased costs for fire and pest control as well as greater investment in health services.

The epistemic community has developed consensus on the buildup of greenhouse gases and threats (as well as promises) of climate change, yet scientists are not unanimous in their predictions of long-term consequences.

A small but influential group of "greenhouse skeptics"[2] repeatedly has advised policymakers to defer action on the climate issue, pending the results of further research. They have argued that even if climate risks are serious, the penalty for a few decades of inaction will be small.[3] They have been especially influential in the United States. However, in nation-states and regions where coastal zones are critical to key economic activities—including tourism, fisheries, agriculture, flood control, and (perhaps ironically) oil and gas extraction and refining—the effects of sea level rise as well as the increased frequency and intensity of storms and storm surges has begun to affect policy making even while scientific uncertainty persists. In these cases local knowledge (the observations and orally transmitted historical records of indigenous and long-time resident populations) has sometimes been used to supplement scientific findings as sources of policy relevant data; and a variety of national and international approaches have been made to mitigating such problems as beach erosion and the degradation of coral reefs, coastal wetlands, and mangroves.[4]

2.2. International Action

Notwithstanding the greenhouse skeptics, the scientific community was active among the governments of industrialized nations and in particular, those in Europe.[5] In 1988 the United Nations formed the Intergovernmental Panel on Climate Change (IPCC), with strong support from the epistemic community.[6] The first IPCC report, issued in 1990, projected an average increase in global temperature of from 1 to 3 degrees Centigrade by the year 2100. Negotiations among nations led to the construction of the Framework Convention on Climate Change (FCCC), which was adopted by the 1992 Earth Summit in Rio de Janeiro.

As Soroos notes, the Framework Convention set the broad terms for negotiation, ultimately leading to the Kyoto Protocol of 1997.[7] The FCCC set a broad goal of reducing human interference with the global climate system. It acknowledged that the economically developed countries bore most of the responsibility for an increase in greenhouse gas (GHG) concentrations. Finally, it called on these countries (called the Annex I countries in the Kyoto Protocol) to reduce GHG emissions to the 1990 levels by the year 2000. It was at the third Conference of the Parties (COP) to the FCCC, held in Kyoto, Japan in 1997, that the Protocol was signed, pledging economically developed countries to reduction targets.[8] To ease American objections to the treaty, negotiators accepted several "flexibility mechanisms," enlarging the means countries could use to meet reduction targets, and "joint implementation" ventures giving credits to EDCs for emission reductions investments they made in LDCs.

The American withdrawal from the Kyoto Protocol in 2001 seemed to jeopardize the Kyoto Protocol, which required ratification by countries emitting 55 percent of the global total of GHG. However, the international legal process concluded in late 2004 when the Russian Duma and President Putin agreed to the accord.

2.2.1. Response of Economically-Developed Countries

Nations of the North are not uniform in their response to the climate change issue. The most active nation in this regard has been Germany, which has led the European Union in the formation of both regional and national policies to reduce greenhouse gas emissions. Particularly notable has been the effort to develop carbon taxes and to endorse by ratification the Kyoto Protocol. The Netherlands and the Scandinavian states also have been active proponents of addressing climate change issues. Within the EU, however, Great Britain and southern European countries such as Italy and Greece have been more reluctant participants in global negotiations. New entrants to the EU, which also are less well developed economically, have emphasized "burden sharing" within a "bubble," where northern, economically developed states make larger sacrifices than southern, developing countries, and EU members are treated as a single entity rather than individual nation-states.[9]

The U.S. position on climate change has been influenced more by the fossil fuel industry than by the epistemic community, notwithstanding important differences in presidential administrations. The administration of the first President Bush (Republican) was a reluctant participant in early international negotiations, including Rio, but the Democratic Clinton administration agreed to stabilization of carbon dioxide emissions at 1990 levels. Then the second Bush administration in 2001 withdrew the United States from the Kyoto Protocol.[10]

As noted previously, the Japanese government initially was not supportive of climate change negotiations. However, it hosted the Kyoto discussions, and then pledged itself to emissions reductions. In sum, the variation among economically developed countries suggests that economics alone explains insufficiently. Differences in interest group structure (especially the power of the fossil fuel industry), strength of NGOs such as Greenpeace within different countries, ideological differences with respect to energy use, different patterns in the conservation of energy, and even separation of powers provide necessary aspects of the explanation.

The Kyoto ratification process illustrates the persistent tensions between international commitments and national motives, interests and institutional factors. In the U.S. ratification of Kyoto became a victim of partisan politics. The agreement, initialed by President Clinton, was first supported then withdrawn from Senate consideration by his Republican successor George W. Bush in the face of apparent opposition by Mr. Bush's close supporters in the energy industry and key Republican senators. Contrast this to a transitional presidential system with a weak legislative branch and an executive that has managed to keep powerful economic interests at bay, politically. For President Putin of Russia legislative and industry opposition were not serious constraints. President Putin faced no effective organized opposition in the Russian Duma, and the substantial deindustrialization suffered by the Russian economy after the 1991 dissolution of the Soviet Union meant that Russia could qualify for valuable carbon credits under Kyoto without changing its current practices or jeopardizing earnings for its energy sectors.

2.2.2. Responses of Less Developed Countries

LDCs also are divided on the issue of climate warming. The subset of nations relying on oil exports for obvious reasons have been opposed to carbon dioxide emissions limits, which would reduce petroleum sales. Thus OPEC nations have played little positive role in climate change negotiations.[11] On the other hand, other nations of the global South, of various sizes and levels of development, have responded to and even taken leading roles in international environmental agreements.

Small island developing states (SIDS) have been particularly vulnerable to climate change and therefore particularly active on the issue. Forty-three states of the Caribbean, Mediterranean, Atlantic, Pacific and Indian Ocean regions have convened a series of conferences on sustainable development under UN auspices. Beginning with the 1994 Barbados conference and the Barbados Programme of Action, SIDS have attempted to generate data, share learning and institutional capacity, and attract and distribute funding from multilateral and bilateral aid agencies for addressing problems of biodiversity conservation, sustainable tourism, renewable energy, coastal and marine resource conservation, as well as climate change. Similarly, the Alliance of Small Island States has seen some success in attracting funding for climate change research and mitigation. Typically, small island states are only lightly industrialized, and with the exception of the few oil exporting states like Trinidad and Tobago and Brunei, contribute very little to the production of greenhouse gases (GHGs). Therefore, they are unlikely to receive direct benefits from carbon credit trading regimes or other market based mechanisms. Instead they must rely on development assistance, technology transfers and offsets from developed states. As climate change threatens beaches, coral reefs and other charismatic landscapes, SIDSnet (the Internet-based information and communication facility of the 43 SIDS conferees), the Alliance of Small Island States, the Organization of Eastern Caribbean States, the South Pacific Regional Environment Programme, *inter alia*, in partnership with local, regional and international NGOs have gained a voice in the climate change debate and attracted funding from international organizations and the governments of developed countries. But the effectiveness of these programs and organizations for addressing the effects of climate change has been limited by domestic factors within each nation-state. Problems include insufficient national capacity to implement the needed policies, inefficient and sometimes corrupt local governments, bureaucratic and partisan differences, and resistance by government agencies and commercial interests (both local and transnational).[12]

Newell suggests that there are three other groupings of developing nations with respect to the climate change issues.[13] First are the largest (in land area and population) nations such as China, India, and Brazil, which emphasize the need for differentiated responses among the LDCs. They have resisted targets for emission levels at present but ask that industrialized nations accept targets as well as provide technology transfers and assistance to LDCs. Some of these states, most notably China, have developed programs for carbon sequestration.

A second grouping, including most of the Group of 77 nation-states, ask that developed nation-states (the Annex I countries) make all the sacrifices at the present and into the future. The final grouping includes countries such as Argentina and Kazakhstan, which have offered to make voluntary emission reductions.

While level of economic development is an important factor in these differentiated responses, so is geographic position and economic focus. At present, political variables appear to be less important in explaining differences, but political development in authoritarian states could change that.

3. BIODIVERSITY LOSS

3.1. Scientific Evidence and Economic Consequences

Biological diversity refers to the variety of living organisms on earth, the range of species, the genetic variability within each species, and the varied characteristics of ecosystems. Today, loss of species and their habitats is a problem of global dimensions; it potentially undermines the equilibrium supporting ecological security. Well over 1,000 species per year may be disappearing, compared to only 1-4 species per year from the fossil record.[14]

It is difficult to understand exactly what impact human activities have on biodiversity, because the total number of species in the world is unknown.

However, a recent study estimates the total number of existing species as about 13 million, of which less than 2 million have been described.[15] Most intensely studied are plants and chordates (fish, birds, mammals). Studies estimate that the impact of human activities on other species has threatened the continued existence of 18 percent of mammals, 11 percent of birds, 8 percent of plants, and 5 percent of fish.[16]

Deforestation is a primary cause of biodiversity loss, as forests are home to more than one-half of all species. Population growth and the timber industry are the primary factors causing a substantial reduction in the world's forests.[17] Biodiversity loss has enormous consequences for humans. In economic terms alone, global threats to species and ecosystems may cost at least $33 trillion.[18] The increasing loss of species threatens purification of air and water, food security, and complex compounds used in medicines, among other adverse consequences. Significantly for developing countries, however, the economic costs of biodiversity loss can be diffuse and are not always felt immediately; while reaping the benefits of biodiversity conservation can require trade-offs with the more immediate and economic benefits of development.

3.2. International Conventions

From the 1970s through the 1990s, countries developed a series of international conventions for the purpose of protecting endangered and threatened species and habitats. The first and most prominent was the 1973 Convention on the International Trade in Endangered Species (CITES), through which countries pledged to ban the importation and exportation of endangered species listed in the CITES annexes. CITES operates by requiring signatory parties (most of the world's nation-states) to regulate international trade in species listed in its appendices. Appendix I includes species threatened with extinction by international trade. These species are strictly regulated and are not allowed to be commercially traded internationally. Appendix II includes all species which may become extinct if their trade is not regulated. To engage in trade for an Appendix II species, a CITES permit is required. Species may be added or deleted from these two restrictive appendices only by a two-thirds majority vote at a Conference of Parties (COP) of CITES.

Other treaties protecting biodiversity include the International Convention for the Regulation of Whaling, the Convention Concerning the Protection of the World Cultural and Natural Heritage and the Ramsar Convention (Convention on Wetlands of International Importance Especially as Waterfowl Habitat). The most comprehensive convention, however, was the 1992 Convention on Biodiversity (CBD).

Like other conventions, the CBD attempted to affect the development choices of member countries. Unlike other conventions, it attempted to protect terrestrial life forms residing principally within the borders of sovereign states. Thus, the purpose of the CBD is not the joint management of a common, global resource, but the "coordinated management of domestic resources" around the globe. At the crux of the CBD are decisions about land use and in particular the setting aside of lands for multiple "non-productive" uses. Because economically developed nations have already devoted most of their natural resources to production, the onus was placed on developing countries to set land aside for the preservation of biodiversity. Another function of the CBD is then to find ways to compensate developing countries for limiting development. An important component of the convention was recognition of aboriginal and community use of biological resources, and emphasis on traditional as well as modern forms of ecosystem knowledge.

3.3. Variation in National Biodiversity Protection Regimes

The range of values attached to biodiversity has expanded and evolved with the movement. Early activism was often associated with movements to preserve "charismatic macro-fauna" such as whales, dolphins, elephants, and pandas; or with the preservation of habitats of great national beauty enshrined in national parks and monuments. More recently, biodiversity has become associated with the preservation of valuable genetic materials from a wide range of life forms, including bacteria, plants, and insects. These changes affect the economic and political calculations of states and communities that control or have access to shrinking repositories of biodiversity. Immediate economic needs may call for the clearing of rain forest, the building of dams, or the planting of modern, high-yield crops. But there may be long-term benefit in the preservation of plants with medicinal properties, insects that control agricultural pests, forests that protect watersheds and downstream water supplies and soils, or traditional crops that provide the genetic bases for nutritionally valuable hybrids. Thus, biodiversity has become a kind of global public good. States face pressures from international organizations as noted above, from NGOs, multi-national corporations, and their own populations to find ways to preserve future value without sacrificing present needs. Tensions arise between the pursuit of immediate, specific value and long-term, diffuse value.

Notwithstanding these common aspects of uncertainty and tension, nation-states vary greatly in the extent to which they have established policies and practices to protect threatened and endangered species and the habitats critical to their existence. In this section we describe and analyze *biodiversity protection regimes*, defined as the complex of authorities (laws, regulations, and policies), protected areas, implementing institutions, and monitoring agencies (including ENGOs). The concrete examples used are drawn from two contrasting cases—the United States and China, and brief references are made later to the experiences of two Caribbean SIDS—Dominica and Grenada.

3.3.1. The Authorities

Most economically developed nations established laws, regulations, and policies to protect endangered and threatened species and their habitats in the 1970s. The United States was the pioneer in the development of such legislation, with the establishment of the National Environmental Policy Act (NEPA) in 1970 (41 U.S.C. S 4332). At the heart of NEPA is a requirement that before any major federal action is taken, significantly affecting the quality of the environment, an environmental impact statement (EIS) must be completed. The EIS must comprehensively examine potential impacts of the action on the environment, and clearly specify alternatives, with their environmental effects. Courts have treated NEPA as procedural legislation and typically have not required federal agencies to produce specific substantive outcomes, so long as alternatives are carefully considered. Yet a series of cases in federal courts have challenged the actions of federal agencies in protecting threatened and endangered species, because alternatives have not been designed to protect critical habitat or because the cumulative effects of successive federal actions have not been considered.

The most significant legislation affecting biodiversity, however, is the Endangered Species Act (ESA), passed by the Congress with virtually no opposition in 1973 (16 U.S.C. 1531). Congress enacted ESA:

> [T]o provide a means whereby ecosystems upon which endangered species and threatened species may be conserved, to provide a program for the conservation of such endangered species and threatened species, and to take such steps as may be appropriate to achieve the purposes of treaties and conventions set forth in this subsection.[19]

ESA outlines a management process to provide for listing and protection of threatened and endangered species, which begins with an individual or group petition to the relevant agency (primarily the Fish & Wildlife Service [U.S. FWS] of the U.S. Interior Department for terrestrial species, and the National Marine Fisheries Service [NMFS] of the U.S. Commerce Department for marine species).

Once a species is listed, the agency organizes a recovery team and develops a recovery plan to outline the potential causes of population decline with recommendations to promote recovery. Section 4(3) of ESA requires that "critical habitat" be designated within one year of the listing, defined as:

> [A] specific geographic area that is essential for the conservation of a threatened or endangered species and that may require special management and protection . . . [It] may include an area that is not currently occupied by a species, but that will be needed for its recovery.[20]

When critical habitat has been designated, more restrictive management regulations are required to reduce adverse impacts to the species.

The most powerful section of ESA is Section 7, which calls for the consultation of all federal agencies to "insure that any action authorized or carried out by such agency is not likely to jeopardize the continued existence of any endangered species ... or result in the destruction or adverse modification of habitat." Should a proposed action, for example, authorize a groundfishery to operate in an area of decline in population of a marine species or the building of a road in a forest with endangered birds or animals, the action must be modified. The agency must consider mitigation alternatives or even abandon the action if *jeopardy* to the species or *adverse modification* to its critical habitat cannot be avoided.

Section 7 provides for a consultation process to define proposed actions regarding the species, identify and involve affected interests, and design attempts to mitigate adverse effects to the species. Significantly, decisions in the consultation process must be based on the "best scientific and commercial information available," and not on the grounds of the economic or other interests affected. The final result of the consultation process is a biological opinion that indicates whether a species is in jeopardy. If the agency makes a jeopardy or adverse modification finding, it must issue a "reasonable and prudent alternative (RPA)," which provides protection for the species, before any federal action may continue. Moreover, the process can be challenged in federal court if a finding of "no jeopardy" is issued.[21]

A third U.S. law, the Administrative Procedures Act (APA) (5 U.S.C. SS 702, 706), outlines the procedures under which actions of administrative agencies, such as the U.S. FWS or NMFS, can be challenged in federal court. Allegations of illegal conduct by agencies customarily are filed in a U.S. district court, and the cognizant judge makes a determination based on the administrative record (AR) of the agency. Although district courts are trial bodies, the determination is based on the record and is not a *de novo* proceedings, for example, with the opportunity for either of the parties to call upon testimony by expert witnesses available for cross-examination. The judicial proceeding is not based on canons of scientific investigation, experimentation, and certainty. Instead, judges ask whether the evidence, meaning the administrative record, does or does not support the complaint of illegal agency action, with interpretation conditioned by relevant judicial precedents. If the agency action does not conform to the judge's interpretation of the law, the action is ruled "arbitrary and capricious," and is remanded to the agency for correction.

Altogether, these authorities are regarded as the world's most strict measures in the protection of biodiversity. NEPA is a model applied extensively throughout other economically developed nations and in many LDCs. The ESA, too, has been widely copied; however, the ability to challenge implementing agencies in national courts depends on the independence of this institution, which varies cross-nationally.

China is a contrasting example of biodiversity protection. From the establishment of the People's Republic of China in 1949 through the Maoist era, little attention was paid to preservation of endangered or threatened species. Indeed, movements associated with the Great Leap Forward and Cultural Revolution were the most dangerous assaults on species and fragile ecosystems in Chinese history.[22] It was only in the reform era under the leadership of Deng Xiaoping in 1978 that environmental issues including biodiversity began to be addressed.

In the 1980s, a series of laws focused on different ecosystem types, with most pertaining to the protection of China's forests. The first forest legislation was enacted by the National People's Congress in 1985. This Forest Law formalized the division of forests between the state and collectives. It enunciated principles for forest management, set up a timber harvest quota system, and required permits for shipping timber. In 1985 the Grassland Law was enacted, and initial regulations were promulgated on nature reserves.

It was not until 1988, however, that species protection legislation—the Wildlife Protection Law—was enacted by the National People's Congress, and it was promulgated the following year. The act lists about 1,300 species as protected under two categories—I and II. Species in both categories are considered threatened, but the type I or "key" species are in greatest need of protection. With the exception of several orchid species, the great majority are mammals. Moreover, the law imposed penalties for killing or trading in banned species. The maximum penalties were quite harsh, including long prison terms and even execution. Sayer and Sun note that "more than 30 people have been executed for killing or trading in parts of elephants and giant pandas."[23]

Additional legislation on the marine environment, fisheries, pollution prevention, water and soil conservation, and land management extended the reach of the state further into protection of degraded ecosystems and species.

And, in the 1990s, several policies and programs emphasized ecosystem protection. In 1998, the National Forest Protection Program established a goal of protecting 61.1 million hectares of forests in the upper reaches of the Yangtze and Yellow Rivers and 33 million hectares in the Northeast and Inner Mongolia. After disastrous floods on the Yangtze in 1998, logging in the middle and upper reaches of the Yangtze and Yellow Rivers was banned.[24] Too, since the 1980s, the state has embarked on an ambitious and aggressive afforestation campaign, as well as a plan to return farmland on steep slopes to forest or grass cover.

3.3.2. Establishment of Protected Areas

The strategy used by most nation-states to preserve species and their critical habitats is to establish nature reserves, parks, forests, refuges, and other restricted-use areas. Classification systems and amount of protection vary by nation-state.

The first protected area was Yellowstone National Park, established by the U.S. Congress in 1872. The greatest majority of protected lands are found on federal public domain, which comprises 29 percent of the total land area of the United States. Initially, public domain comprised 80 percent of American lands, but until the turn of the twentieth century, the Congress sold or gave lands to spur development of infrastructure (such as the trans-continental railroad system), agriculture, natural resource and other economic development activities, and population settlement.

Approximately half of the federal public domain can be considered protected for the purpose of preserving species, habitats, and eco-systems. The American system is composed of five components:[25]

- National Wilderness Preservation System, established in 1964 and including about 106 million acres (with 50 million acres of Alaska wilderness added by the Congress in 1980). These lands by legislative mandate are to remain undeveloped areas forever;
- National Wildlife Refuge System, composed of more than 93 million acres, and including 500 refuges, which provide habitat for migratory birds and animals;
- National Forests, which include more than 190 million acres, and are protected in order to provide timber supplies for future national needs as well as to protect mountain watersheds;
- National Park System, including more than 83 million acres, with 66 national parks and 318 national monuments, historic sites, recreational areas, near-wilderness, seashores, and lake shores, and restricted from mining, logging, and grazing; and

- National Rangelands, including 403 million acres of grassland, prairie land, desert, scrub forest, and other open space, much of which is suitable for grazing.

The Chinese system, on the other hand, has a different basis in organization. Since the establishment of the People's Republic in 1949, all land in China has been owned by the state. With the onset of economic reform in 1978, however, some lands have been managed by collectives, with leasehold rights to individuals. And the Chinese state has addressed biodiversity conservation by establishing nature reserves, forest reserves, parks, and other protected areas. The first nature reserve was established in 1956, but until the end of the Maoist era, few areas received this type of protection. Then, in the 1980s and 1990s, protected areas grew rapidly. By 2005, over 2,000 protected areas had been formed in China; some are quite small but a few comprise large areas of the lands in a province or autonomous region.[26] Altogether, they comprise about 15 percent of China's land area. However, critics allege that from one-third to half of the protected areas are "paper parks."

The first mention of protected areas in planning documents was in 1979. Regulations were promulgated for them in 1985. Revisions to these rules were endorsed by the State Council in 1994. The management and financing of protected areas are controversial and currently under study.[27] By regulation, the protected areas include three separate management zones:

> *[C]ore area* with no use, habitation or interference permitted, not even scientific research; *buffer zone* where some collection, measurements, management and scientific research is permitted; and *experimental zone* where scientific experimentation, public education, survey, tourism and raising of rare and endangered species are permitted.[28]

According to the Protected Areas Task Force Report of 2004, the zoning system copies that proposed by UNESCO for use in Biosphere Reserves, and is designed to allow study of interactions between human use and nature.

3.3.3. Implementing Agencies

In the United States, a number of federal agencies implement provisions of NEPA and ESA regarding species and habitat protection. As mentioned, the U.S. Fish & Wildlife Service has responsibilities for the protection of most listed terrestrial species. Although the area of USFWS operations extends to 152 million acres, it is not a land-use agency with a mission to administer public domain under the doctrine of multiple use.

Instead, it regulates the development, protection, rearing, and stocking of wildlife resources and their habitats. It protects migratory and game birds, fish and wildlife. Also, it enforces regulations for hunters of migratory waterfowl and preserves wetlands as natural habitats. The agency is part of the U.S. Department of the Interior and has requested a budget of $1.3 billion for FY05, which would be an increase of $22.6 million over FY05.

The National Marine Fisheries Service (NMFS) is an agency of the National Oceanic and Atmospheric Administration within the U.S. Department of Commerce. As mentioned, it is responsible for most marine endangered and threatened species and habitats. Under the Outer Continental Shelf (OCS) Act of 1953, the Congress declared federal government ownership of OCS lands extending as far as 200 miles offshore, which comprises an area of more than 1 billion acres. NMFS differs from USFWS in that it has a dual mission, to enhance the American fisheries under national fisheries legislation such as the Magnuson-Stevens Fisheries Conservation Act and the American Fisheries Act, and to protect fisheries and other marine resources. Within the agency two distinct offices—the Office of Sustainable Fisheries and the Office of Protected Resources—perform these functions. NMFS is a relatively small agency with approximately 2,500 total staff nationally and a requested FY 2006 budget of $727.9 million.

The U.S. Forest Service has a jurisdiction extending over 191 million acres. It regulates the use of forest resources and the activities of commercial foresters working in national forests. This is among the oldest conservation agencies in the United States, established in 1905 as part of the U.S. Department of Agriculture and founded by Gifford Pinchot, one of America's leading conservationists. As Rosenbaum notes, "with more than 38,000 employees and a budget exceeding $2.0 billion, the Forest Service historically has possessed a strong sense of mission and high professional standards."[29] Nevertheless, it was the Forest Service that was brought to its knees by the Spotted Owl controversy in the 1980s, and required by federal court order to curtail logging in old growth forests of the Pacific Northwest.

Like the Forest Service, the Bureau of Land Management has multiple responsibilities. In addition to management of wildlife habitats and endangered plant and animal species, it also manages timber, minerals, oil and gas, geothermal energy, rangeland vegetation, recreation areas, and wild and scenic rivers. It is an agency of the U.S. Interior Department and directly manages 264 million acres of public domain (and leases another 200 million acres in national forest and private lands).[30] Although the lands administered are far greater than those of the Forest Service, BLM's budget and staff are less than a third of this agency. It also lacks the autonomy that the older agency has developed, and is subject to the political currents of the presidential administration in power.

The National Parks Service also is an agency of the U.S. Department of the Interior, with responsibility for 83 million acres. Its mission is to conserve the scenery, natural and historic objects, and wildlife in the nation's parks. Two other agencies have responsibilities for endangered and threatened species and habitats, but largely in the context of NEPA. The U.S. Environmental Protection Agency (EPA) is always a participating agency in the development of environmental impact statements. The Army Corps of Engineers of the U.S. Department of Defense has explicit responsibilities to protect the shorelines of oceans and lakes, and invariably is the lead agency on environmental impact statements for construction projects in navigable waterways.

Compared to the American system of agencies implementing biodiversity protection requirements, the Chinese pattern is equally complex.[31] The organization of China's national bureaucracy is not conducive to effective implementation of biodiversity conservation laws and policies. China developed an environmental agency, the National Environmental Protection Administration, only in 1988. The name of the agency was changed to the State Environmental Protection Administration (SEPA) in 1998, and it was then elevated to ministerial status, but it remains a relatively small agency, with only one-twentieth the personnel of the U.S. EPA yet with a far broader brief.[32] SEPA has broad responsibility for each of China's major environmental problems, and thus its attention is diffused to issues of air, water, and land pollution, acid rain, and climate change. In the area of biodiversity conservation, it has a department of nature conservation and division of nature reserves and species management; it also has general responsibility for developing and maintaining the biodiversity data management and information system.

Because most of China's endangered and threatened species are located in forested areas, the State Forestry Administration (SFA) has broad administrative responsibilities for their preservation. SFA also is a relatively small agency, with only around 250 Beijing office employees. Since the reorganization of central government agencies in 1998, it has enjoyed sub-ministerial status, a notch below the SEPA. SFA is the primary agency for implementation of the Wildlife Protection Act and the Forestry Act, and it is responsible for the management of about 75 percent of China's protected areas. A third agency involved in biodiversity conservation is the Ministry of Agriculture (MOA). Once, this department was housed together with forestry, but they have been independent since 1998. MOA has an office for endangered and threatened species; it administers a small number of protected areas, about 3 percent of the total. Often the jurisdictions of MOA and SFA conflict, as the former is in charge of terrestrial aquatic species, which may be found in protected forestry areas administered by SFA.

A fourth agency is the Ministry of Construction (MOC), which is in charge of China's pubic construction, including national highways, dams such as the Three Gorges Dam, and ports and harbors. The activities of this agency may directly impair threatened and endangered species, which explains its involvement in the biodiversity conservation regime. In addition, it administers a small number of protected areas, about 3 percent of the total. The fifth agency is the State Oceans Administration (SOA), whose responsibility extends offshore China's coasts to the 200-mile limit. SOA has responsibility for marine reserves and all endangered and threatened marine species, but identification of such species is least well advanced among the categories of protected species. It administers a small number of marine protected areas.

A number of other agencies are involved in biodiversity conservation less directly. For example, the Ministry of Foreign Affairs (MOFA) is responsible for international conventions in which China participates, and heads Chinese delegations at international environmental conferences. Similarly, commerce and trade offices have duties with respect to certain environmental treaties, such as CITES. Altogether, at least nine central government ministries have some duties in biodiversity conservation; none has a clear lead agency role in all areas of conservation, as also is the case in the United States. Nevertheless, some integration is achieved through the formation of task forces and working groups, frequently coordinated under the State Council.

A much noted tendency of Chinese government since the onset of economic reform is devolution of administrative power to provinces and autonomous regions, and this practice vastly complicated biodiversity conservation efforts. For example, each of the provinces has a forestry administration office, and forestry bureaus are found at the municipal level too. The sub-national offices operate in a problematic administrative context as they serve two masters: the SFA in Beijing and the provincial governor (or local mayor). Because administrative control tends to follow funding and the national government allocates less to environmental conservation than provinces and municipalities, there is no clear line of authority from the center to the site where problems of endangered and threatened species conservation must be resolved. Provincial and local environmental offices are relatively well supplied with personnel—from 60,000 to 120,000 in the early twenty-first century.[33]

3.3.4. Monitoring by ENGOs

Most of the mainstream environmental organizations in the United States direct some attention to endangered species and habitat issues. Perhaps most influential are the National Wildlife Federation, the Sierra Club, the National Audubon Society, the Wilderness Society, Friends of the Earth, the Environmental Defense Fund (now called Environmental Defense), and the Natural Resources Defense Council. As Rosenbaum notes, these organizations are "thoroughly professionalized and sophisticated in their staffs and organization and are armed with the same high-technology tools and modern techniques of policy advocacy as any other powerful national lobby."[34] Several have membership rolls numbering in excess of 500,000 members.

Non-mainstream organizations such as Greenpeace also have played roles in biodiversity protection work. Greenpeace maintains its reputation as a radical organization by direct action, such as sending representatives on boats to disrupt trawl fishing in the North Pacific or disrupt whaling. Yet it also has joined with other ENGOs in suing implementing agencies for their failure to protect endangered species and their habitat. For example, in 1998, Greenpeace joined the Sierra Club and American Oceans Campaign in petitioning the U.S. District Court of Western Washington (in Seattle) to force the NMFS to protect the endangered Steller sea lion and its habitat in the Gulf of Alaska, Bering Sea and Aleutians Islands. The court, finding jeopardy to the species and adverse modification to its critical habitat, briefly closed the $1 billion groundfish fishery until the agency had developed RPAs to protect the species.[35]

The monitoring activities of ENGOs in economically developed nations are well-known. Within the last two decades, they have made appearances in many LDCs. ENGOs are relatively new agents in Chinese society, and because China remains an authoritarian state system, the role of NGOs is weak. Particularly after the student demonstrations at Tiananmen in 1989 and the Falun Gong protests, the regime has scrutinized NGOs carefully. It permits only those promoting state objectives, and to the present has favored ENGOs with foreign connections, as a conduit to international funding. Nevertheless, the rise of environmentalism in China has received important state support. For example, in April 2004 Premier Wen Jiabao suspended plans for a massive dam system on the Nu river in western China that scientists warned would ruin one of the country's last unspoiled places. In a written instruction, Wen ordered officials to conduct a major review of the hydropower project, saying "[W]e should carefully consider and make a scientific decision about major hydroelectric projects like this that have aroused a high level of concern in society, and with which the environmental protection side disagrees."[36]

Of the approximately 2,000 NGOs in China today, only three or four dozen are bona fide NGOs that specialize in environmental protection.[37] Most of the NGOs are what are called GONGOs, or government-organized NGOs (see chapter 3). The small number of NGOs is explained by the sensitivity of the regime to potential dissent and by the onerous registration requirements. Effectively, a group must find a government agency to sponsor its work, and then is limited in its ability to enlist members and raise money locally. Although there are a number of grassroots ENGOs,[38] most are outposts of international ENGOs and located in Beijing. We present brief vignettes of three ENGOS currently active in different areas of biodiversity conservation.

Greenpeace entered Hong Kong in 1997, with a strategy of building up a localized team and then extending its operations to other areas in China. Only in 2002 did Greenpeace establish a program office in Beijing. Two of its objectives pertain to biodiversity. First, more than other organizations in China, it has focused on genetically-modified organisms (GMOs) and biosafety issues. The initial area of emphasis has been on soya where, given China's role as the center of global soybean cultivation, it fears contamination upon introduction of GMOs. Additional emphasis areas include Bt cotton and rice. Second, Greenpeace is examining the international trade in timber of Chinese companies. The organization also publishes newsletters and reports, sponsors scientific conferences, and works with agencies such as SFA to strengthen biosafety protocols.[39]

A second Beijing ENGO established in 1994 is the International Fund for Animal Welfare (IFAW), with a mission to protect wild animals and to promote their welfare. IFAW has been active in monitoring the protection of endangered species, such as the Asian elephant, in protected areas in Yunnan province, and has worked on the development of a trans-boundary elephant park between China and Laos. It has supported government programs against poaching Tibetan antelopes in the Qinghai-Tibet plateau and assisted research and programs to close down bear farms for bile extraction. The organization also has focused on logging that threatens endangered and threatened species, and has been successful both in negotiations with companies to reduce their logging footprint and with government officials to tighten monitoring and enforcement of conservation laws.[40]

A third ENGO is the World Wide Fund for Nature (WWF). It has been involved in conservation activities in China for nearly 25 years, but its Beijing office was not established until 1996. (Most of its earlier work was coordinated from Hong Kong.) WWF is China's largest ENGO, with an office staff of 20 and a 2002 budget of around $400,000. It began its work in China in 1979 by assisting the government in the establishment of the Wolong Giant Panda Reserve. The panda is WWF's logo in China, and it continues panda preservation efforts in the Minshan region, in the Qinling Panda Focal Project (Shaanxi province) and supports surveys and studies on pandas and their habitats. It has also been active in restoration of wetlands on the Yangtze, development of sustainable ecotourism projects, and wetlands conservation in

Tibet.[41] In 2001 the organization launched a China species preservation program, emphasizing protection of less well-known species. It gives \$5,000 (RMB) grants (about US\$617) for investigative and species preservation work.[42]

This sample is sufficient to detect a pattern. The ENGOs all engage in environmental education, to spread knowledge about endangered and threatened species. They tend to be project-specific, providing assistance to government agencies in the preservation of individual species (usually charismatic fauna) and specific ecosystems or threatened eco-regions. Most of the organizations have limited local memberships. They are highly reliant on international headquarters or other offices for financial support of operations in China. Finally, they are all moderate in their approach to the government and search for the most effective tools to use in the Chinese context. Said one ENGO coordinator, "Some actions we might take would be counter-productive, for example, blocking the driveway to the premier's house. We are not protesting for the sake of protesting."[43]

3.4. Challenges to Effective Biodiversity Conservation

Five specific problems challenge conservation of threatened and endangered species on a global basis: the legal framework; horizontal and vertical administration; financial resources and incentives; human resources and training; and value conflicts. These challenges affect all countries but are particularly serious in economically developing countries. For this reason, most of the discussion below concerns China, a mega-diversity country with problems in species preservation like those of most LDCs.

3.4.1. The Legal Framework

The framework of laws, regulations, and policies for the conservation of species is relatively comprehensive in the United States. And the U.S., like most EDCs, is a society where the "rule of law" prevails, notwithstanding differences in national administrations and variations among state and local jurisdictions. Yet many critics have alleged that the focus on single species is misguided and the legal framework needs to encompass ecosystems. A different aspect is the controversial nature of the ESA in the United States, and the failure of the Congress to re-authorize it, since its authorization expired in 1992.

Although China has a large number of laws on species and ecosystem preservation, there are areas of overlap (perhaps unavoidable) and serious gaps. Moreover, most of the laws are vague and ambiguous, reading more like policy statements than directives, making administration and enforcement difficult. The large system of nature reserves is administered on the basis of regulations and not law. The regulations do not conform to international standards, such as those of the World Conservation Union, on categories of protection. The zonation system especially imposes hardships on people in rural areas, who may be displaced or whose livelihoods may be ruined, without adequate compensation. Because of the serious attention the State Council has given recommendations of the China Council for International Cooperation on Environment and Development (CCICED) in the past, one can be optimistic that China will shortly enact comprehensive legislation on protected areas.

Other gaps in the skein of legislation may be more difficult to resolve. Species preservation legislation and regulations say little about protection of plant species and nothing about small and economically insignificant species such as insects. Coral species are omitted too.

A final legal framework issue is the lack of connection in law between species and their critical habitat. This is a particular problem for migratory species, which might be sheltered in one stopping point on a migration route but not others. Protection of ecosystems serving as critical habitat to large numbers of species is difficult in any country, and it is difficult to be optimistic about its success in China.

Interestingly, some problems similar to those noted above for China have recently resurfaced in the U.S. Changes in enforcement practices by the U.S. EPA pertaining to the Clean Air and Clean Water Acts, and attempts by Congress to remove habitat protection from the purview of the Endangered Species Act also leave gaps in policy implementation.[44] However, unlike the Chinese case these gaps and ambiguities are attributable to ideological and partisan conflict and the changing impact of interest group activity on presidential administrations and congressional leadership. Whereas, in the U.S., current changes are attributable to (and future changes will likely result from) electoral turnover in the executive and legislative branches, in China change could require some basic reconsideration of development strategies.

3.4.2. Administrative Organization and Enforcement

In previous sections we have alluded to horizontal integration problems in the United States regarding species and critical habitat preservation. Many federal agencies, in at least four different departments, have protective responsibilities, and there is no designated agency that plays the lead on all biodiversity cases and issues. Then within agencies, conflicts appear to the extent that missions conflict. For example, the NMFS has primary responsibilities for fisheries enhancement in the exclusive economic zone (EEZ) of the United States; it also has protective responsibilities for endangered marine species, such as the Steller sea lion, and its initial failure to protect this species against jeopardy and the adverse modification of its critical habitat landed the agency in federal court for five years.[45]

The ESA effectively nationalized species and habitat protection in the U.S., putting federal officials in charge and overriding state and local authorities. Nevertheless, most state wildlife departments list species of concern and set aside protected areas for their preservation. Federal and state activities in species preservation, however, are not coordinated.

The problems of administrative organization in China are both horizontal and vertical. At the central government level, protective functions are divided among a large number of agencies, each with different missions. A forestry administrator gave an example of this type of conflict:

> For example, the alligator is an endangered species in China . . . Yet today, there is conflict between the SFA, which believes it should manage the species as it is a land species, and the MOA, which believes it is in charge of reptiles and amphibians using the water system. (Why isn't the conflict resolved, perhaps by the State Council?) The State Council is absorbed with political problems and issues of economic development. It doesn't have the time to resolve a matter such as which agency should be in charge of which species. Then there is conflict with SEPA, which has authority regarding biodiversity preservation.[46]

The Task Force report calls for greater data sharing and collaboration among agencies, which might reduce this conflict. However, without designating lead agencies for single or groups of species, or ecosystems, the conflict seems unlikely to be resolved given the insulated behavior of national ministries in China.

The problem of vertical integration is much more intractable, and parallels the difficulty China has faced in coordinating economic development activities from the center to provinces and municipalities. The problem here extends beyond the system of divided loyalties, with provincial and municipal environmental bureaus responding to both national and sub-national masters. It also reflects the difficulty of having different systems of incentives and values. One NGO representative who had visited 60 protected areas explained the conflict between national conservation objectives and local practices as follows:

> There are many different local situations, and especially in the remote areas (where the largest number of protected areas

> are). The biggest factor is the drive for rapid economic development. Local governments want to develop the economy. And they want to measure the efficiency of their officials by economic development and not their conservation efforts. So they look at economic development needs first. Each local government administration has only 4-5 years to get promotions, and they focus on economic development. It is short-term, non-sustainable economic development, and that's the main stress to the environment in local areas. Local governments and companies want to build big dams. They earn money from the construction of dams. Such projects bring money to the local people, yet they are often harmful to the environment.[47]

National ministries retain supervisory authority. For example, an official of the SFA said that in June 2004 some 35 officials from his department would fan out over China to check implementation of the Wildlife Protection Law and the Forest Law. Upon their return, they would file reports with the administrator, who in turn would report on implementation to the State Council.[48] While he believed the reports would be candid and forthright, the officials would not be able to inspect more than a few nature or forest reserves in each province/autonomous area, and would lack sufficient information to make charges of maladministration were they to find it. In fact, few formal complaints have been made by state ministries against provincial or local environmental protection bureaus in their species conservation work. As one forestry official remarked: "Different levels of government have different interests; but our government structure is unitary, and it assumes that everyone will share the same interest. This is delusion of thought."[49]

3.4.3. Financial Resources and Incentives

In the United States, resources allocated to biodiversity protection at the federal level have increased significantly from the 1970s through the early twenty-first century, yet the growth has been erratic because of frequent budget cuts, particularly in the Reagan and G.W. Bush administrations. Moreover, congressional critics of the ESA have held budgets of some protective agencies hostage, altered agency funding priorities, and forced reduction in personnel. Nevertheless, federal salaries are sufficiently high to attract good personnel.

Several financial problems beset China's biodiversity conservation efforts. First, China has become overly reliant on foreign funding for management and training, which is unsustainable beyond the near-term. Second, insufficient funding has been allocated to species and ecosystem preservation efforts. One estimate is that only $100 million RMB (about US$12.3 million) recently has been allocated from the central government for protected areas, which has to be divided among more than 2,000 reserves.[50] When protected areas developed most rapidly in the 1980s and 1990s, there were incentives for their establishment as the central government allocated one-time-only funding for this purpose. Operation and maintenance funding, however, depends on the classification of the protected area. National-level reserves were funded wholly by the central government; provincial and municipal-level reserves were funded primarily by sub-national governments. These provincial and local governments tended to spend little on reserves, and one consequence of this development was an increase in the number of national-level reserves (about 225 today).

Cash-starved managers of provincial and municipal-level protected areas then created their own incentives. Those areas with charismatic fauna or other aesthetic values emphasized eco-tourism. Those areas with exploitable natural resources began to develop them. This increased revenues for administration of the reserves but obviously had deleterious effects on the eco-system and species being protected.

The unsettled "ownership" status of many rural ecosystems complicates the resource allocation problem. Although all land in China is owned by the state, only 60 percent or so of nature and forest reserves are on land over which state agencies have clear control. The remainder are controlled by collectives (*jiti*), operating under the "responsibility system," initiated at the onset of economic reforms. For example, Harkness notes:

> When the collective forest lands of Yuhu village were incorporated into the Yulongxueshan Nature Reserve in north-west Yunnan . . . farmers responded by cutting down trees they had previously managed on a sustainable basis.[51]

The ownership status of collectives remains unclear, but may be clarified as property rights become better defined in China. Nonetheless, the central government's funding of reserves on collectives as well as their supervision remain problematical.

3.4.4. Human Resources and Training

In the United States, the biological and ecological sciences are well-developed, and universities annually produce sufficient trained graduates to staff federal protective agencies. Too, university and research institute scientists interact frequently with agency protective personnel.

The human resource issues in China are typical of developing countries globally. At the national level, knowledge of individual species and ecosystems is limited. The biological sciences are poorly developed in China, and taxonomy is particularly underdeveloped. This adversely affects the development of comprehensive data bases on biodiversity.[52] Ecology as a discipline is barely two decades old. Although China has benefited from foreign expertise, this too is localized to specific regions and species. China today lacks the knowledge base to support a sustained campaign to protect endangered and threatened species.

Officials in central government ministries are likely to have been trained domestically, but in institutes supporting the mission of their agencies. For SFA, for example, this means staff are likely to have graduated from the six universities which specialized in forestry. However, the emphasis of these educational programs was on forest development for commercial purposes, and not forest ecosystem protection. An optimistic note is the increased number of officials who have been educated abroad or who have benefited from training institutes and programs.

At the provincial and local levels, the preparation of officials for conservation of endangered species and ecosystems has been abysmal. Until recently, appointments to positions were made on patronage grounds. They lacked any knowledge of the life sciences, could not distinguish the species they were hired to protect, and had no incentive to increase their knowledge or capability. Also, as one official noted, in the remote, rural, isolated PAs, "[T]he conditions for staff are difficult. To stay there, essentially, they have to sacrifice their children's future and sometimes their marriages."[53]

3.4.5. Value Conflicts

In the United States, the main obstacle to biodiversity conservation is economic: the value of the critical habitat in which rare species reside for other purposes—fisheries, logging, recreation, housing, agriculture, hydropower development, mining and oil and gas extraction. Conflicts between preservation and economic development values have produced controversy, such as virulent opposition of the wood products industry, loggers, and Pacific Northwest communities to the closure of old growth forests to logging and opposition of the $1 billion annual groundfish fishery to closure of zones in the Gulf of Alaska and Bering Strait/Aleutian Islands in order to protect the Steller sea lion.

Three conflicts in values obstruct the strategy of biodiversity conservation in China.[54] First and uppermost is the conflict between seeking rapid economic development, to make China a "middle class" country by 2020 (as announced by Jiang Zemin in 2002) and the sustainable development of China, which implies the preservation of its threatened and endangered species and ecosystems. It is pressures for rapid economic development which led to both economic and environmental devolution. In general, China's richer areas (primarily those along the east coast) have allocated more resources to species conservation with greater success. In China's poorer regions, where most of the protected areas are located, preservation of species conflicts with provincial and local attempts to foster economic development. Harkness argues that "Conservation remains a largely unfunded mandate even inside the nature reserve system, with fiscal pressures leading some reserve managers to cannibalize the very resources they are supposed to protect."[55] In the competition between hungry humans and threatened non-human species, the former have won out consistently. As one official opined, "If people don't have enough to eat, you can't expect them to protect the environment."[56]

A second value conflict is between the central direction of environmental policy and local initiatives and participation. Most of the biodiversity conservation efforts in China have emphasized national strategies, and have not involved local communities. Large numbers of people have been displaced from their homes and communities to serve conservation values; those remaining constantly face threats to their livelihoods. There is little consensus in local areas of China that biological diversity should take precedence over a variety of uses of natural resources.

The third value conflict expresses the difference in culture and lifestyle of Han Chinese and non-Han minority populations who may approach conservation in unorthodox ways, and who populate many protected areas, especially those on China's periphery. For example, Harris suggests that "Wildlife communities and habitats are generally healthier and more intact in areas of predominantly ethnic minority occupancy than in ethnic Han areas."[57]

Interestingly, the opposite seems to be true in the U.S. While economic interests seem to be more important than values for explaining conflict over biodiversity policy, economic cleavages do frequently coincide with regional, ethnic, racial, socio-economic and cultural cleavages. The conflict over opening Alaska's Arctic National Wildlife Refuge to oil exploration, for example, not only pits pro-development forces against environmentalists, but the oil based economic interests of one native group against the economic and cultural values attached to subsistence hunting by another. Also in Alaska, local proponents of wolf control policies aimed at increasing moose populations for both subsistence and recreational hunters frequently characterize their opponents as "Outside" environmental radicals and urbanites unfamiliar and unconcerned with the needs of "true Alaskans."[58] Conflicts over habitat protection, such as the previously mentioned spotted owl controversy or the recent show-down between federal officials and farmers in

Oregon over water rights and salmon habitat, take on cultural as well as economic dimensions. And elsewhere in the United States, less developed areas with higher percentages of ethnic minorities do seem to suffer more from the effects of environmental degradation, leading to charges of environmental racism.[59]

3.5. Challenges to Small States

As mentioned in our discussion of climate change, small states face particular challenges. This is also true of biodiversity. Even with small and sometimes declining populations the pressures of economic development, deforestation, desertification and other forms of habitat destruction can endanger critical species.

In the small island states of the Eastern Caribbean political structures and processes involved in the making and implementation of biodiversity policy and the adjudication of disputes tend to be simpler, even more poorly funded and more open to outside influence than in either the United States or China. As parliamentary democracies with often fractious party systems, Caribbean SIDS are subject to sudden shifts in policy and turnovers of top ministerial personnel due to elections and cabinet reshuffling. Responsibility for economic development and environmental policy are often spread among ministries and agencies with conflicting missions related to species protection and resource development. Ministerial portfolios and cabinet posts are frequently combined, separated and recombined to deal with shortages of capable personnel and the changing priorities of governments.

In Dominica, commercial crops, watershed management, fisheries, marine habitat, coastal zone management, forestry, and terrestrial wildlife policy are all the responsibility of divisions of the Ministry of Agriculture and the Environment. Therefore, marine and forest reserves, parks and protected areas must compete for limited resources and staff support with agencies administering agricultural export policy and commercial fisheries. And one or two officers will be responsible for regulation of sport fisheries, small scale local commercial fishing, marine protected areas, recreational boating and dive operations, whaling and marine ecotourism. Budgets are chronically short, making departments dependent on outside sources—including international organizations, multilateral lending institutions, foreign governments and ENGOs—for the financial and technical resources, salaries and training needed to comply with international environmental agreements and conventions.[60] But external funding has tended to be insufficient and not always efficiently used, and can carry with it the stigma of outside interference, which can be a political liability.[61] In 2002, for example, a legislative stalemate resulting from an unstable coalition government stalled plans by the

Ministry of Agriculture and the Environment to use GEF funds—received for implementation of provisions of the UN FCCC—to create an Environmental Coordinating Unit to oversee the ministry's development and resource management activities.

Local and international ENGOs and international organizations have been active and occasionally influential in Dominica and in the rest of the Eastern Caribbean. The World Wildlife Fund—with U.S. Agency for International Development (USAID) funding—implemented a highly successful coastal zone clean up and conservation campaign in the 1990s. In 1997, Dominica's Morne Trois Piton National Park was declared a World Heritage Site by UNESCO, and in 1998 the president of the Dominica Conservation Association received the Goldman prize for leading a campaign that convinced parliament to reverse its decision to allow an Australian mining multinational to explore for copper in pristine rainforest.[62]

On the other hand, the Japanese government has been a generous donor in the development of commercial fisheries in Dominica and actively courts the support of the Dominican government and other Eastern Caribbean governments in support of its positions in the International Whaling Commission. Japan's desire to pursue commercial whaling in the Eastern Caribbean, with its year round populations of sperm whales, bottle-nosed and spinner dolphins, and migratory populations of humpback and pilot whales, has put the Dominican government at odds with local and international ENGOs and local ecotourism operators.[63]

In other small island developing states, however, external pressure and funding have facilitated the development of institutional capacity for biodiversity conservation. In the late 1990s in Grenada funding from the World Bank, Caribbean Development Bank and GEF supported a process, facilitated by a regional ENGO (the Caribbean Natural Resources Institute), that established a new national park to preserve habitat for the Grenada Dove—a decidedly uncharismatic but highly endangered species. The creation of the park also set in train a series of events that would lead to a new Forestry Policy, supported by a grant and technical support from the British Department for International Development. The Forestry Policy process involved advanced training at British universities for forestry officers, the formation of a new divisional structure for the Forestry Department (including a division of biodiversity conservation), and the organization of user groups. The process also helped the conservation-minded forestry officers to overcome cabinet-level political resistance and realize their longstanding ambition to wrest control of the Department of National Parks from the more development oriented Ministry of Tourism.[64]

4. NATIONAL "STYLES" IN RESPONSE TO GLOBAL ENVIRONMENTAL CHANGE

Nation-states vary in how they respond domestically and internationally to evidence of global environmental crises. The explanations for these differences are complex, but the cases presented above suggest some general patterns. Responses are conditioned by: the domestic institutional capacity already in place, and the inclinations both of external and domestic political actors to make more resources available; the economic interests in play and the levels of development of states and societies; the economic incentives made available for compliance with or resistance to international agreements and conventions; and the basic characteristics of national political systems.

4.1. Institutional Capacity and Economic Competitiveness

Ultimately, the geographical and resource attributes of developing countries may be more important than political factors in determining their participation in international agreements and the development of domestic measures for climate change mitigation and biodiversity protection. Economically developing countries that export fossil fuels (such as OPEC members and Mexico), and those undergoing rapid transitions to highly industrialized, middle income status (such as China and India), may be expected to resist external pressures and/or demand special consideration under international agreements such as the Kyoto Protocol. But an effectively assertive response to external pressures—whether that response is to cooperate or resist—depends on substantial internal strength.

The existence of trained personnel (including significant local participants in global epistemic communities), bureaucratic structures and resources, budgetary support for the formulation and implementation of regulatory and sustainable development policy, will affect both the recognition of and response to crisis. In the United States, with its extensive and multilayered regulatory mechanisms, advanced research capacity and strong, institutionalized linkages between government agencies and academic researchers, environmental information is readily received and processed by governments. Conflicts between environmental and economic development policy do result in shifting budgetary priorities, as well as significant variations in the style, aggressiveness and targeting of environmental policy enforcement. But regardless of who holds the White House, environmental interests will invariably get a hearing in Congress and the courts, and environmental policy will remain an important part of the mission of several departments, agencies and commissions. Furthermore, shifts in policy and implementation strategies will occasion strong reactions by members of Congress, national and international ENGOs, and other interested publics.

In China, as is typical of many authoritarian states (and especially post-revolutionary regimes) development has taken precedence. In an approach to environmental issues that is ironically close to neoliberalism, radical Maoist development policies followed a strategy of "grow first, clean up later."[65] In an effort to stimulate rapid aggregate growth and modernization in the post-Mao period, scarce state resources have favored an environmentally compromising combination of big infrastructural projects (such as the Three Gorges Dam), and reliance on cheap energy (especially soft coal) for rapid industrial expansion. Nevertheless, along with economic reforms has come greater attention to the environmental impacts of development. Still, regulatory and enforcement structures are poorly staffed and have ambiguous or even contradictory mandates. Retaining some of the key limiting factors of an economically developing country, despite its growing wealth and power, China continues to rely on foreign assistance and ENGOs for addressing problems of air and water quality, GHG production, and biodiversity conservation.

For poorer, smaller nation-states compliance with international agreements on biodiversity and climate change depends greatly on external support (grants, low interest loans, technology transfers, training, etc.) from multilateral and bilateral aid agencies and ENGOs. These states must contend with poverty, a paucity of high-value marketable products, limited human resources, and daunting challenges from climate change and species extinction. Government agencies charged with the implementation of the UNFCCC and biodiversity conventions suffer from inadequate budgets and political weakness. Without strong linkages to business interests, and/or powerful political patrons (domestic or foreign) they rarely prevail in their contests for influence with agencies that facilitate industrialization, resource extraction, conventional tourism and similar routes to increased foreign investment and short term economic growth. Where species protection and climate change mitigation policies are effectively implemented it is often as a result of rent-seeking behavior by governments looking for new sources of foreign aid and concessionary loans, rather than sustained commitments to environmental improvement.

4.2. Democracy, Dictatorship and the Environment

There is reason to expect that democratic regions will respond more effectively to environmental crisis.[66] As discussed in chapters 2 and 3, democracy allows freedom for social movements and NGOs to operate legally, and opens multiple points of access to policy makers. In addition, green movements can spawn green parties. Democratic regimes are also typically middle and upper income countries and therefore more likely to take a longer view of resource-related issues and find more support for policies that further postmaterial values (see chapter 2).

But while the cases described above do not invalidate the positive relationship of political democracy and economic development to environmentalism they do show that democracy is not always a good predictor of the specific responses of particular governments. The existence of "rule of law" in the United States does not eliminate the substantial bureaucratic discretion available to executive branch agencies, and does not rule out dramatic shifts in policies and commitments from one presidential administration to another. Even in developed countries displaying the economic and political attributes typically associated with compliance with international treaties and the promulgation of effective domestic regulations, ideological and partisan conflict can cause significant variability in the biodiversity and climate change programs over time.

In China, one consistent feature of economic development policy—from Maoist radicalism to market reforms—has been a strong emphasis on growth. Democratization, decentralization and greater openness to foreign influences may eventually bring value change to economic policy-making. The effectiveness of ENGOs in campaigns to protect "charismatic megafauna," like the giant panda, suggest a set of circumstances in which the political interests of national and local leaders, dominant economic interests and global pressures may converge to produce effective biodiversity conservation.

The experiences of the microstates of the Eastern Caribbean show that democracy alone is not sufficient to allow positive responses to environmental crises. In fact, by opening policy-making and implementation to competitive pressures from a broad array of domestic and external economic and political influences, democracy in poor, highly dependent states may pressure governments to favor short term economic interests over longer term environmental interests as policy-making becomes a contest between the proponents of conservation and growth.

---

[1] Gunther Weller and Patricia Anderson, eds., *Implications of Global Climate Change in Alaska and the Bering Sea Regions*. Fairbanks, AK: Proceedings of a Workshop, Center for Global Change and Arctic System Research, University of Alaska Fairbanks, 1998. See also the Arctic Research Consortium of the United States (ARCUS), *People and the Arctic: A Prospectus for Research on the Human Dimensions of the Arctic System*. Fairbanks, May 1997; National Science and Technology Council (NSTC), *Our Changing Planet: The FY2000 U.S. Global Change Research Program*. Washington, DC: A report by the Subcommittee on Global Change Research, Committee on Environment and Natural Resources, 2000; National Assessments Synthesis Team, *Climate Change Impacts on the United States*. Cambridge: Cambridge University Press, 2000.

---

[2] For example, the Marshall Institute, *Scientific Perspectives on the Greenhouse Problem*. Washington, DC: George C. Marshall Institute, 1989; S. Fred Singer, ed., *Global Climate Change: Natural and Human Influences*. New York: Paragon House Publishers, 1989; Richard Kerr, "Greenhouse Skeptics Out in the Cold," *Science*, Vol. 146 (December 1, 1989): 1118-19; and Richard S. Lindzen, "A Skeptic Speaks Out," *EPA Journal*, Vol. 16 (March-April, 1990), 46.

[3] M.E. Schlesinger and X. Jiang, "Revised Projection of Future Greenhouse Warming," *Nature*, Vol. 350 (1991), 219-21.

[4] One example is Caribbean Planning for Adaptation to Climate Change, supported by the Global Environment Facility, World Bank, and Organization of American States; see http://www.unitedcaribbean.com/cpacc.html.

[5] Clair Gough and Simon Shackley, "The respectable politics of climate change: the epistemic communities and NGOs," *International Affairs*, Vol. 77, No. 2 (2001): 329-45.

[6] See Peter Haas, "Epistemic Communities and International Policy Coordination," *International Organization*, Vol. 46, No. 1 (winter 1992), 1-35.

[7] See Marvin S. Soroos, "Negotiating our Climate," in Sharon L. Spray & Karen L. McGlothlin, *Global Climate Change*. Lanham, MD: Rowman & Littlefield Publishers, Inc., 2002, 126-27.

[8] Member of the European Union agreed to reductions of GHG emissions by 8 percent by 2012; the United States, under the Clinton presidency, agreed to 7 percent reductions; and Japan and Canada agreed to 6 percent.

[9] See Peter Newell, *Climate for Change: Non-state Actors and the Global Politics of the Greenhouse*. Cambridge: Cambridge University Press, 2000, 14.

[10] For an explanation of U.S. domestic policy formation on climate change, see Aaron M. McCright and Riley E. Dunlay, "Defeating Kyoto: The Conservative Movement's Impact on U.S. Climate Change Policy," *Social Problems*, Vol. 50, No. 3 (2003): 348-73.

[11] See G. Porter and J. Brown, *Global Environmental Politics*. Boulder, CO: Westview Press, 1991.

[12] See, http://www.sidsnet.org/; http://www.sidsnet.org/pacific/sprep/; and Jonathan Rosenberg and Linus Spencer Thomas, "Participating or Just Talking? Sustainable Development Councils and the Implementation of Agenda 21," Global Environmental Politics, Vol. 5, No. 2 (May 2005): 61-87.

[13] Newell, 2000, 17-18.

[14] Tim O'Riordan and Susanne Stoll-Kleemann, *Biodiversity, Sustainability and Human Communities*. Cambridge: Cambridge University Press, 2002, 14. Estimates of the number of species disappearing vary greatly.

[15] United Nations Environmental Program (UNEP), *Global Biodiversity Assessment*. Cambridge: Cambridge University Press, 1995.

[16] Dennis Pirages and Theresa DeGreest, *Ecological Security*. New York: Rowman & Littlefield, 2004.

[17] Jerry McBeath, "Perspectives on Deforestation," in *Global Environmental Politics*, Vol. 3, No. 3, 108.

[18] Stuart Pimm et al., "The Value of Everything," *Nature* (May 15, 1997).

[19] The full text of ESA can be found at: http://www.law.cornell.edu/uscode/16/1531.html.

[20] Ibid.

[21] Section 11(g)(1)(1) gives legal standing to any person who "may commence a civil suit on his own behalf . . . to enjoin any person, including the United States . . . who is alleged to be in violation" of the ESA.

[22] See Judith Shapiro's *Mao's War Against Nature: Politics and the Environment in Revolutionary China*. Cambridge: Cambridge University Press, 2001.

[23] Jeffrey A. Sayer and Changjin Sun, "Impacts of Policy Reforms on Forest Environment and Biodiversity," in William F. Hyde, Brian Belcher, and Jintao Xu, eds., *China's Forests: Global Lessons from Market Reforms*. Washington, DC: Resources for the Future, 2003, 181.
[24] Sayer and Sun, 181.
[25] Based on the analysis found in Walter A. Rosenbaum, *Environmental Politics and Policy*, 5th edition. Washington, DC: CQ Press, 2002, 311.
[26] For example, nature reserves in Tibet comprise 26 percent of the land. See Wu Ning, Daniel Miller, Lu Zhi, Jimmy Springer, eds., *Tibet's Biodiversity: Conservation and Management*. Beijing: China forestry Publishing House, 2000, 9.
[27] See "Protected Area Task Force (PATF) Report to CCICED 2004: Evaluation on and Policy Recommendations to the Protected Area System of China," Draft, April 2004. For a review of China's protected areas, see Xie Yan, "Review of China's Natural Conservation Area Management System," in *China Environment and Development Review* (in Chinese), Beijing: Social Sciences Documentation Publishing House, 2004, 273-95. For the final report, see Xie Yan, Wang Sung, and Peter Schei, *China's Protected Areas*. Beijing: Tsinghua University Press, 2005.
[28] PATF Report to CCICED, 2004, 12.
[29] Rosenbaum, 313.
[30] Rosenbaum, 313.
[31] See Jerry McBeath and Jenifer Huang McBeath, "Biodiversity Conservation in China: Policies and Practice," *Journal of International Wildlife Law and Policy* (forthcoming).
[32] Elizabeth Economy, *The River Runs Black: The Environmental Challenge to China's Future*. Ithaca, NY: Cornell University Press, 2004, 263.
[33] This is Economy's estimate. See, 183.
[34] Rosenbaum, 34.
[35] See *Greenpeace v. National Marine Fisheries Service*
[36] Jim Yardley, "China's Premier Orders Halt to a Dam Project Threatening a Lost Eden," *New York Times*, April 9, 2004.
[37] Schwartz estimated that only 40 ENGOs were operating in his 2004 analysis "Environmental NGOs in China: Roles and Limits," *Pacific Affairs*, Vol. 77, No. 1, 36. This corresponds with the author's survey in spring 2004, which identified about 36 viable ENGOs.
[38] For a view of a successful grassroots group, the Upper Yangtze Organization, see J. Marc Foggin, "Highland Encounters," a case study of the *Innovative Communities Initiatives*, September 2003.
[39] Interview with Greenpeace campaign director, Beijing, June 23, 2004.
[40] Interview with director of IFAW, Beijing, June 30, 2004.
[41] See WWF, *'01-03 WWF China Programme Report,* Beijing, 2003.
[42] Interview with communications officer, WWF, Beijing, July 1, 2004.
[43] Interview with ENGO representative, Beijing, June 15, 2004.
[44] Critical commentaries and appeals to members by environmental NGOs, Sierra Club and Clean Water Action, are typical of the responses generated by critics of the Bush (II) Administration's enforcement policies: http://www.sierraclub.org/cleanwater/sewage/; http://www.cleanwateraction.org/.
[45] See J. McBeath, "Management of the commons for biodiversity: Lessons from the North Pacific, *Marine Policy* 28 (2004), 523-539.
[46] Interview with SFA official, Beijing, May 27, 2004.
[47] Interview with ENGO representative, Beijing, June 11, 2004.
[48] Interview with SFA official, Beijing, May 10, 2004.
[49] Interview with SFA official, Beijing, May 17, 2004.
[50] Estimate of SFA official, Beijing, July 8, 2004.

[51] James Harkness, "Recent Trends in Forestry and Biodiversity in China," in Richard Edmonds, ed., Managing the Chinese Environment. New York: Oxford University Pres, 2000, 921.
[52] See Peter Schei, Wang Sung, Xie Yan, compilers, *Conserving China's Biodiversity (II), 1997-2001,* CCICED, Beijing, China Environmental Science Press, 2001, 10-11.
[53] Interview with forestry official, Beijing, July 5, 2004.
[54] For a sensitive appraisal of western criticism of wildlife conservation in China, see Richard Harris, "Approaches to Conserving Vulnerable Wildlife in China," *Environmental Values*, 5 (1996), 303-34.
[55] Harkness, 2000, 929.
[56] Interview with SEPA official, Beijing, May 30, 2004.
[57] Harris, 1995-6, 10.
[58] For example, Bob Robb, "Alaska's wolf-control crises," *The American Hunter*, May 2003.
[59] See, for example, Kelly Michele Colquette, and Elizabeth A. Henry Robertson. "Environmental Racism: The Causes, Consequences, and Commendations." *Tulane Environmental Law Journal 5* (1): 153- 207, December 1, 1991.
[60] Interviews with Fisheries Officers and country representative of the Organization of American States, Roseau Dominica, June 26, 1998 and June 7, 1999.
[61] Interviews with an officer of the Dominica National Sustainable Development Council, Roseau, Dominica, May 21, 1998, May 31, 1999 and February 2002.
[62] Interview with the President of the Dominica Conservation Association, May 21, 1998; and http://www.goldmanprize.org/recipients/recipients.html.
[63] Interviews with the President and Programme Director of Dominica Conservation Association, Roseau, Dominica, May 21, 1998 and June 8, 1999.
[64] Steve Bass, Policy that Works for Forests and People Series No: 10, Participation in the Caribbean: A Review of Grenada's Forest Policy Process. London: International Institute for Environment and Development, 2000; Jonathan Rosenberg and Fae L. Korsmo, "Local Participation, International Politics, and the Environment: The World Bank and the Grenada Dove," Journal of Environmental Management Vol. 62 (July 2001): 283-300.
[65] Vinod Thomas and Tamara Belt, "Growth and the Environment: Allies or Foes?" *Finance and Development,* Vol. 34, No. 2, (June 1997), 22.
[66] Martin Janicke, "Democracy as a Condition for Environmental Success: Non-institutional Factors," in *Democracy and the Environment: Problems and Prospects,* edited by William M. Lafferty and James Meadowcroft. Cheltenham, UK: Edward Elgar, 1996.

CHAPTER 7. SUMMARY AND CONCLUSIONS

We conclude this volume in two sections. First, we summarize the argument of the early chapters. Then we explore the way that comparative politics helps aid understanding of complex environmental problems nationally and internationally.

1. SUMMARY

Chapter 1 introduced the subject of comparative environmental politics, first through contrasting examples of how nation-states have responded to ecosystem crises. It also introduces the environmental problems with global ramifications, specifically climate warming, biodiversity loss, deforestation, desertification, transboundary air pollution, and marine pollution and over-fishing. Then, it explains the subject matter of comparative environmental politics and differentiates it from recent studies in the globalization and international relations literature, which tend to marginalize the role played by domestic social forces and the political structure of nation-states. For three reasons, we emphasize the importance of states as the focus of analysis: (1) they make the critical decisions affecting the global environment, (2) they alone can decide whether to participate in international environmental agreements, and (3) domestic circumstances powerfully influence environmental policies of all nation-states and environmental conditions everywhere. We reviewed differences among nation-states in size, socio-cultural integration and level of economic development as well as by regime type, summarized in Table 1.1. We highlighted the growing number of comparative studies of environmental politics, on which much of this volume is based.

In chapter 2, we turned to the large topic of state-society relations, asking broadly how social and cultural forces have influenced state decision-making. The first substantive section of this chapter treated traditional attitudes and values toward the environment. Beginning with western values, which express the "dominant social paradigm" of human exploitation of nature, the discussion then contrasted Asian, Islamic, African, and aboriginal belief systems concerning the environment. The next section explored patterns of economic activities in states, as categorized by important post-World War II social science concepts of modernization, political development, uneven development, and dependency. The third section asked whether a "new" environmental paradigm is in the process of being formed. Starting from a basis in environmental philosophy, this section discusses post-materialism as part of the broad value shift represented in postmodernism. Survey research documents support for a new environmental paradigm, and in nations outside the West too. However, advocacy of postmaterialist values is greater in economically developed countries (EDCs) than in lesser developed countries (LDCs). In the former, research suggests the operation of an environmental Kuznets curve and ecological modernization, both of which argue that environmental effects of economic development are positive. Yet even in wealthy nations, changes in attitudes toward nature do not add up to a robust definition of sustainable development. This chapter's final section portrays the relationships of social group (including business) organizations to environmental policy. Reviewing studies of pluralism, corporatism and their variants, we found that corporatist structures tended to be more conducive to environmental regulation of production in democratic EDCs. In authoritarian and transitional LDCs, institutionalized relationships between social groups and the state also have powerful effects on environmental policy. But based on a limited number of case studies, we can only tentatively conclude that in LDCs corporatism favors embedded interests that historically have been unsympathetic to environmental groups.

Chapter 3 then explored political processes and organizations. The environmental non-governmental organization (ENGO) originated in western societies of the post World War II era and has expanded to most nation-states today. ENGOs vary cross-nationally in scope, membership, leadership, purpose, linkages to governments, and orientations to the system, as cases from EDCs and LDCs, large countries and small states demonstrate. The second section of this chapter considered the intimate relationship between ENGOs and national environmental movements, and then treated ecological resistance movements—more likely to be found outside liberal societies and to disproportionately represent racial/ethnic minorities and women. Three cases feature ecological resistance: the Ogoni of Nigeria, rubber tappers in the Brazilian Amazon, and broad-scale mobilization against a large dam project in India. The third section examined the origin of Green parties, the role electoral institutions have played in facilitating or retarding their development, their activities, and still limited electoral appeal. Next, this chapter explored roles of media and public opinion in forming environmental consciousness and driving policy outcomes. The chapter concluded with analysis of the linkage between environmentalism and democratization, which varies as a causal relationship by country and level of economic development. Even in already liberal states, environmentalism may have the effect of increasing participatory democracy.

Chapter 4 focused on political institutions, beginning with an explanation of their nature, and treated four dimensions. First mentioned were constitutional limitations on state power and differences between liberal and authoritarian systems. In the former, environmentalism is much farther advanced. Second, the chapter treated territorial distributions of authority, contrasting federal and unitary systems. Because of extensive devolution of power from unitary states to sub-national units, this comparison is skewed. Yet federalism does appear to make a difference in environmental outcomes when combined with other factors such as electoral institutions, political culture, and corporatist structures. Third, we analyzed degrees of concentration of power, contrasting presidential, parliamentary, and corporatist systems, and finding that deconcentrated systems tolerated a larger number of ENGOs. Fourth, the chapter examined the special role courts played, if independent, in environmental decisions of nations. Then, the chapter reviewed the ways in which institutional components interacted in the policy-making process and the political opportunity structure. In the former area, we analyzed the special characteristics of environmental problems, the methods used to establish majority coalitions, and the stages of the environmental policy process. In the latter area, we highlighted the importance of political institutions as they influence environmental interests, parties, and movements.

Chapter 5 introduced the subject of national capacity to protect the environment, discussed who governed environmental outcomes, and described what resources—economic, human, and political—are required to build capacity. Central to environmental capacity is administrative competence of states, for example whether or not they have centralized and effective national environmental ministries. Also of importance has been long-term, strategic environmental planning. The chapter then roughly measured environmental capacity by sorting nations into three categories. A small number of states, all EDCs, were at one point or another "pioneers," including the United States and Great Britain. Another small subset of nations has attained the status of "models," such as Germany, the Netherlands, and Sweden. Most nations, however, are neither: they encounter serious "implementation deficits" in managing environmental policy and are "laggards," as seen in the cases of China, Nigeria, and Russia. The chapter noted some exceptions to generalizations on implementation deficits, for example in small island states which have established sustainable development councils to overcome capacity problems. In general, case studies demonstrated a strong correlation between per capita GDP and environmental indicators. Too, cases showed the generally positive impact of horizontal diffusion from pioneers and models to environmental laggards, and vertical diffusion from international NGOs, UN environmental agencies, and global lending agencies.

Chapter 6 connected domestic to international dimensions of environmental politics, through analysis of national responses to environmental problems. The first issue treated was climate change, and we discussed the scientific background to the issue and the relevant international conventions. Then we contrasted the way EDCs responded to the issue as compared to LDCs. The second issue, treated in greater detail, was biodiversity loss. Again we examined the scientific evidence and economic consequences of threats to species and destruction of their critical habitat. Then we analyzed variation in national biodiversity protection regimes. We compared the authorities (laws, regulations, and politics) to protect species and ecosystems, showed how protected areas were established in the United States and China, presented information on implementing agencies, and surveyed their monitoring by ENGOs. Then, the chapter evaluated five specific problems challenging conservation of species and ecosystems: the legal framework, horizontal and vertical administration, financial resources and incentives, human resources and training, and value conflicts. Because these challenges are greater in LDCs than EDCs, our prime example was China, a mega-diversity country. As an aside we outlined the particular challenges faced by the small island states. The chapter concluded with analysis of national "styles" in response to global environmental change. These tended to vary by institutional capacity and economic competitiveness as well as by the basic characteristics of the political system.

2. THE MERITS OF COMPARATIVE POLITICAL ANALYSIS

Comparative political analysis seeks to understand decisions by tracing the interplay of values, institutions, behaviors and processes within the historical, socio-cultural and economic contexts established by nations-states. By tradition, comparative politics confines itself to the national and sub-national levels. Interactions among states are reserved for the sub-fields of international relations and international political economy. But these sub-disciplinary boundaries have never been perfectly clear. Trespassing is common and trespassers have produced important new insights. In international relations, the levels of analysis debate (see chapter 1) visits and revisits the national and sub-national roots of state behavior. Proponents of the modernization and dependency schools argue over the roles played by global and domestic actors in determining the types and levels of development found within nation-states (see chapter 2). Fernando Enrique Cardoso and Peter Evans, to name just two, combined studies of coalition formation among national political and economic elites with analyses of the international capitalist economy to explain development and its limits in dependent states.[1] Scholars of comparative politics have never felt obligated to limit their studies to policies and behaviors without international effects or implications. Indeed, it is hard to imagine what those might be. Appropriately, textbooks for comparative politics courses at colleges and universities now almost invariably include discussions of globalization,[2] and political scientists writting about globalization freely integrate societal and systemic levels of analysis.[3]

The tools of the comparativist—large number cross-national studies, most-similar and least-similar-case comparisons, and theoretically informed qualitative case studies—have yielded a rich body of observations and propositions on the fundamental questions of politics. Over time, the shifting and competing paradigms of comparative political analysis have improved our understanding of the formation, stability and functionality of institutions; the requisites and prerequisites of democracy and dictatorship; the causal chains that link culture, values, attitudes, and behavior; and the complex interactions among economic factors, social structure and political power.[4]

Environmental politics, because of the transboundary nature of most environmental problems and the global concerns raised by the fate of charismatic creatures and landscapes, provides rigorous tests for the saliency of comparative political analysis. We have selected sets of examples—countries and cases—that suggest how those tests may be conducted. What we have done is quite preliminary, because at the present stage of knowledge about global environmental problems, that is all that can be done. As mentioned in chapter 1, more systematic, cross-national comparisons will require more accurate indicators of environmental conditions—pollution levels, degree of endangerment to species and ecosystems—for most countries on the planet. But the studies cited in this volume indicate that comparative political analyses have already begun to identify and hypothesize about variations in the propensity of nation-states for addressing environmental problems within their own borders and globally.

Looking at climate change, we address an issue that is intrinsically global, and has spawned high-profile international regimes, agreements and epistemic communities. Yet we are struck by the amount of national variation in the search for solutions by states, and the importance of national-level decision-making in determining the success or failure of international environmental regimes. To explain Russian policy on the Kyoto Protocols, for example, we need to understand the relationship between executive and legislative branches and the current state of the Russian economy. To explain the different responses of the EU and United States we look to differences in electoral systems, patterns of interest group representation and the concentration of political power within and among national and sub-national governments.

The study of biodiversity loss makes an equally compelling case for comparative political analysis. International campaigns and networks have been critical for creating marine protected areas, nature reserves, and parks in LDCs and EDCs that are home to endangered species. But the actual level of protection achieved ultimately rests on the capacity of states to make and implement effective policy. Lack of capacity and competing domestic political interests make "paper parks" an all too common problem in LDCs.

Yet EDCs that cooperate actively in international regimes such as CITES still experience intense political battles over habitat preservation and predator control within their own borders, and fail to satisfy international standards on biodiversity protection. One cannot understand the spotted owl controversy in the U.S. state of Oregon, or the determination of Alaska's state government to continue aerial wolf hunting, without understanding the economic, cultural and ideological cleavages in those two states and the relationship between the states and the federal government.

Environmental politics tends to confirm a consistent finding of comparative political analysis: the persistence of systematic differences between developed and developing countries in policy-making capacity, development strategies, political openness, rule of law, and the efficacy of civil society organizations. But it also raises questions that require additional research. For example, comparing Chinese and U.S. ENGOs, environmental bureaucracies and legal practices reinforces some of what we already believe about the differences between authoritarian and democratic regimes, and developed and developing economies. Yet it is not entirely clear that authoritarian polities lack the ability to attain progressive environmental outcomes. China's afforestation and reforestation campaigns are the largest in human history, with the potential for making major contributions to carbon sequestration. And Cuba has begun to abandon large-scale sugar cultivation and is pursuing aggressive programs in organic agriculture and the cultivation of naturopathic medicines. And comparing the effects of corporatism on environmental groups in Western Europe and Mexico demonstrates the need to refine propositions about interest groups and social policy at the national level. Corporatist OECD countries tend to perform better than pluralist ones in environmental protection. But party corporatism in Mexico protects entrenched interests with authoritarian political "styles" making it difficult for environmental groups to enter "normal" politics even as Mexico democratizes.

An important question for any study of environmental politics is whether the environment constitutes a politically unique issue area that invalidates conventional disciplinary and subdisciplinary approaches to political science. We observe in chapter 1 that globalization poses significant challenges to the ability of states to protect and promote the interests of citizens, and that environmental issues, agreements and networks play an important role in political globalization. But we also present numerous examples of the "point source" nature of global environmental problems to show that the cause and cure lay within the sovereign territory of individual states. This can be true of air and water-borne pollution, global warming, and especially biodiversity loss.

Extinction of the lowly Grenada Dove or the spectacular Giant Panda would be losses to the global environmental heritage, and efforts to protect them have taken on international dimensions. But the governments of

Grenada and China made the policies that allowed these international efforts to proceed. In Grenada it was local activitists who provided the World Bank with information that compelled it to pressure the government to save the dove. In China panda lovers solicited international interest in its fate, helped stir up global protest, brought INGOs to China, and instructed them on how to deal with the government. In both cases international networks required domestic political actors to achieve desired outcomes.

Ecological modernization theorists argue that environmental issues have changed the basic fault lines of domestic political competition. In EDCs, they contend, environmentalism creates new divisions among social groups, forcing alliances across class lines, uniting owners and workers in older, polluting industries against the onslaught of postmaterialist values. Mol also argues that these new bases for political competition are spreading globally, overcoming differences in levels of economic development among nation-states and generating value change in LDCs as well as EDCs.[5] Yet, where Green political parties are established features of the political landscape—principally in EDCs with parliamentary systems—they are almost invariably situated on the left, drawing support from middle-class youth, progressive intellectuals, and professionals in the service and postindustrial sectors. The picture in LDCs is less clear. In countries transitioning from authoritarianism, environmentalists may make common cause with worker and peasant organizations on the left. Plantations and factories exploit workers *and* damage the environment. But these alliances may be based more on a mutual desire for change than agreement on a new ideological paradigm. Mexico's green party started on the left and then experimented with an electoral alliance with a conservative party, winning congressional seats only because of a modified proportional representation rule. Brazil's workers' party became nationally competitive with the assistance of grassroots environmentalists. In office, the worker's party has defended peasants' property rights and made moves against illegal logging, but has yet to establish a convincing record on the environment. In short, these developments beg further analysis using the tools of comparative political research.

Our analysis of policy-making too underscores the value of comparative political analysis. Environmental issues have qualities that set them apart, particularly the scientific uncertainty and irreversibility of problems such as habitat destruction, resource depletion, and species extinction. Additionally, nation-states have created institutions specifically dedicated to environmental policy-making. But environmental issues may also be seen as complex public goods problems—along with national defense, public order and infrastructural development—which, it has long been been argued, are at the heart of the formation and function of states and governments.[6] States weigh environmental protection against the other expectations of their citizens. Government officials may defy international and domestic pressures, even when it results in a loss of financial support—as in the case of the Narmada dam projects in India—to pursue projects they consider vital for economic development and the legitimacy of the state. In the name of environmental values, activists take risks to their personal wellbeing that seem to exceed any possible direct benefit to themselves. Some rational choice theorists have suggested that a desire for public goods (rather than personal fulfillment) may motivate rebellious collective action.[7] Others argue that policy (including the provision of collective benefits) is best understood as a private good produced by state officials in exchange for a base of public support.[8] It will be worthwhile to discover the part that environmental policy plays in these decisions (or exchanges), including how and why nation-states differ in this regard.

Global networks have been important features of environmental politics, especially in LDCs where they generate financial and technical support not available locally. A combination of domestic and international political forces—newly organized and occasionally effective civil society actors—has made environmental issues a new and compelling set of policy problems for states. But the long-term survival of green movements and parties depends on an array of domestic political factors that will either impede or allow their inclusion in "normal politics."

Environmentalists have been partners in political change, but nowhere have environmentalists alone (acting globally or locally) affected structural change in the power relationships between states and their societies. Janicke et al. show that while institutional capacity is positively associated with policy innovations, nothing about the changes in environmental policy-making capacity in LDCs has challenged or substantially altered approaches to problem-solving by states. In newly industrializing LDCs, increasingly effective approaches to "end of pipe" pollution exist alongside rising levels of pollution and resource degradation from rapid economic development. In general, democratic EDCs are more effective in addressing environmental problems.[9] But the occasional exceptions to this rule—the Eastern Caribbean cases mentioned in chapters 3, 4, 5 and 6, for example—point to the need for a clearer understanding of variations in national capacity for environmental protection.

Finally, there is a reflexive benefit to a comparative politics of the environment. Comparative political analysis itself should benefit from a continuing examination of environmental issues. The newness of the issues and the institutional responses of nation-states help break down persistent barriers to fully comparative work. For example, in chapter 2 we have chosen to consider western attitudes and values toward the environment as traditional, rejecting the teleology and misleading dichotomies of modernization theory. As a practical matter, effective environmental policy will need to overcome the conventionally "modern" and look for its guideposts in some combination of new approaches and established traditions. Studies on postmaterialism use cross-national comparisons to correlate the desire for new approaches to economic growth and social development with postmodern values (at least in EDCs); but these studies do not tell us what these new approaches might be or how to find them. Work on the integration of scientific and local or indigenous knowledge—employing systematic analyses of the politics of national development, resource management, social policy and indigenous peoples—take us farther in the desired direction.

Finally, studies of environmental politics tend to be less parochial than traditional comparative political studies. Most of the multi-author volumes cited in this book are international efforts. And the strict division between the developed and developing worlds that still permeates comparative politics as a formal sub-field is less evident in comparative environmental politics. Although there appear to be more cross-national studies of developed countries than of developing countries and an insufficient number of large-N studies of either, there are a substantial number of volumes that treat both.

The three environmental crises described at the beginning of this volume were caused by human activity and negligence. Each case exposed insufficiencies of national and local governance, and each case generated political responses. The "hope" in that section's heading, however, does not come from the effectiveness of those responses. New institutions developed, laws changed, public awareness increased, victims were indemnified, and perpetrators were punished. But none of the situations has been fully resolved and victims continue to struggle with the consequences. Instead, hope comes from the fact that three very different nation-states did respond, that the responses were different—in degree, kind and effectiveness—and that by studying the reasons for those differences we may learn to do better.

---

[1] F.H. Cardoso, "Associated-Dependent Development: Theoretical and Practical Implications." In Alfred Stepan, ed. *Authoritarian Brazil*. New Haven: Yale University Press, 1973; Peter Evans. *Dependent Development: the Alliance of Multinational, State, and Local Capital in Brazil*. Princeton: Princeton University Press, 1979.

[2] See, for example, Jeffrey Kopstein and Mark Lichbach, eds. *Comparative Politics: Interests, Identities, and Institutions in a Changing Global Order*. Cambridge and New York: Cambridge University Press, 2000; and Michael Curtis, ed, *Introduction to Comparative Government, 5th edition, update*. New York: Longman, 2006.

[3] The examples are numerous. Perhaps the best known is Benjamin Barber. Jihad vs. McWorld. New York: Times Books, 1995.

[4] David Collier; James E. Mahon, Jr. "Conceptual 'Stretching' Revisited: Adapting Categories in Comparative Analysis." *The American Political Science Review*, Vol. 87, No. 4. (December, 1993): 845-855. Peter Evans; John D. Stephens. "Studying Development since the Sixties: the Emergence of a New Political Economy." *Theory and Society*, Vol. 17, No. 5, Special Issue on Breaking Boundaries: Social Theory and the Sixties. (September, 1988), pp. 713-745.

[5] Arthur P.J. Mol, *Globalization and Environmental Reform: the Ecological Modernization of the Global Economy*. Cambridge, MA: the MIT Press, 2001.

[6] Albert O. Hirschman. *Exit, Vocie and Loyalty*. Cambridge MA: Harvard University Press, 1970; Charles P. Kindleberger. "International Public Goods without International Government. *The American Economic Review*, Vol. 76, No. 1 (March 1986): 1-13.

[7] Edward N. Muller and Karl-Dieter Opp. “Rational Choice and Rebellious Collective Action.” *The American Political Science Review*, Vol. 80, No. 2. (Jun., 1986): 471-488.
[8] Anthony Downs. *An Economic Theory of Democracy*. New York: Harper, 1957. Albert Breton. “A Theory of the Demand for Public Goods.” *The Canadian Journal of Economics and Political Science*, Vol. 32, No. 4 (November 1966): 455-67.
[9] Martin Janicke, “The Political System’s Capacity for Environmental Policy: The Framework for Comparison,” in Helmut Weidner and Martin Janicke, eds, *Capacity Building in National Environmental Policy*. Berlin: Springer, 2002, 409-10.

INDEX

Advances in Global Change Research

1. P. Martens and J. Rotmans (eds.): *Climate Change: An Integrated Perspective.* 1999 ISBN 0-7923-5996-8
2. A. Gillespie and W.C.G. Burns (eds.): *Climate Change in the South Pacific: Impacts and Responses in Australia, New Zealand, and Small Island States.* 2000 ISBN 0-7923-6077-X
3. J.L. Innes, M. Beniston and M.M. Verstraete (eds.): *Biomass Burning and Its Inter-Relationships with the Climate Systems.* 2000 ISBN 0-7923-6107-5
4. M.M. Verstraete, M. Menenti and J. Peltoniemi (eds.): *Observing Land from Space: Science, Customers and Technology.* 2000 ISBN 0-7923-6503-8
5. T. Skodvin: *Structure and Agent in the Scientific Diplomacy of Climate Change.* An Empirical Case Study of Science-Policy Interaction in the Intergovernmental Panel on Climate Change. 2000 ISBN 0-7923-6637-9
6. S. McLaren and D. Kniveton: *Linking Climate Change to Land Surface Change.* 2000 ISBN 0-7923-6638-7
7. M. Beniston and M.M. Verstraete (eds.): *Remote Sensing and Climate Modeling: Synergies and Limitations.* 2001 ISBN 0-7923-6801-0
8. E. Jochem, J. Sathaye and D. Bouille (eds.): *Society, Behaviour, and Climate Change Mitigation.* 2000 ISBN 0-7923-6802-9
9. G. Visconti, M. Beniston, E.D. Iannorelli and D. Barba (eds.): *Global Change and Protected Areas.* 2001 ISBN 0-7923-6818-1
10. M. Beniston (ed.): *Climatic Change: Implications for the Hydrological Cycle and for Water Management.* 2002 ISBN 1-4020-0444-3
11. N.H. Ravindranath and J.A. Sathaye: *Climatic Change and Developing Countries.* 2002 ISBN 1-4020-0104-5; Pb 1-4020-0771-X
12. E.O. Odada and D.O. Olaga: *The East African Great Lakes: Limnology, Palaeolimnology and Biodiversity.* 2002 ISBN 1-4020-0772-8
13. F.S. Marzano and G. Visconti: *Remote Sensing of Atmosphere and Ocean from Space: Models, Instruments and Techniques.* 2002 ISBN 1-4020-0943-7
14. F.-K. Holtmeier: *Mountain Timberlines.* Ecology, Patchiness, and Dynamics. 2003 ISBN 1-4020-1356-6
15. H.F. Diaz (ed.): *Climate Variability and Change in High Elevation Regions: Past, Present & Future.* 2003 ISBN 1-4020-1386-8
16. H.F. Diaz and B.J. Morehouse (eds.): *Climate and Water: Transboundary Challenges in the Americas.* 2003 ISBN 1-4020-1529-1
17. A.V. Parisi, J. Sabburg and M.G. Kimlin: *Scattered and Filtered Solar UV Measurements.* 2004 ISBN 1-4020-1819-3
18. C. Granier, P. Artaxo and C.E. Reeves (eds.): *Emissions of Atmospheric Trace Compounds.* 2004 ISBN 1-4020-2166-6
19. M. Beniston: *Climatic Change and its Impacts.* An Overview Focusing on Switzerland. 2004 ISBN 1-4020-2345-6
20. J.D. Unruh, M.S. Krol and N. Kliot (eds.): *Environmental Change and its Implications for Population Migration.* 2004 ISBN 1-4020-2868-7
21. H.F. Diaz and R.S. Bradley (eds.): *The Hadley Circulation: Present, Past and Future.* 2004 ISBN 1-4020-2943-8
22. A. Haurie and L. Viguier (eds.): *The Coupling of Climate and Economic Dynamics.* Essays on Integrated Assessment. 2005 ISBN 1-4020-3424-5

Advances in Global Change Research

23. U.M. Huber, H.K.M. Burgmann and M.A. Reasoner (eds.): *Global Change and Mountain Regions.* An Overview of Current Knowledge. 2005 ISBN 1-4020-3506-3
24. A.A. Chukhlantsev (eds.): *Microwave Radiometry of Vegetation Canopies.* 2006 ISBN 1-4020-4681-2
25. J.Mc Beath and J. Rosenberg: *Comparative Environmental Politics.* 2006 ISBN 1-4020-4762-2